ALL APES GREAT AND SMALL

VOLUME 1: AFRICAN APES

Developments in Primatology: Progress and Prospects

Series Editor: Russell H. Tuttle, University of Chicago, Chicago, Illinois

This peer-reviewed book series will meld the facts of organic diversity with the continuity of the evolutionary process. The volumes in this series will exemplify the diversity of theoretical perspectives and methodological approaches currently employed by primatologists and physical anthropologists. Specific coverage includes: primate behavior in natural habitats and captive settings; primate ecology and conservation; functional morphology and developmental biology of primates; primate systematics; genetic and phenotypic differences among living primates; and paleoprimatology.

All Apes Great and Small
Volume 1: African Apes
Co-edited by Biruté M. F. Galdikas, Nancy Erickson Briggs, Lori K. Sheeran, Gary L. Shapiro, and Jane Goodall

ALL APES GREAT AND SMALL

VOLUME 1: AFRICAN APES

Co-Edited by

Biruté M. F. Galdikas

President
Orangutan Foundation International
Los Angeles, California

Nancy Erickson Briggs

California State University at Long Beach
Long Beach, California

Lori K. Sheeran

California State University at Fullerton
Fullerton, California

Gary L. Shapiro

Vice-President
Orangutan Foundation International
Los Angeles, California

and

Jane Goodall

President
The Jane Goodall Institute

Springer Science+Business Media, LLC

ISBN 978-1-4757-8236-3   ISBN 978-0-306-47461-3 (eBook)
DOI 10.1007/978-0-306-47461-3

©2001 Springer Science+Business Media New York
Originally published by Kluwer Academic/Plenum Publishers, New York in 2001
Softcover reprint of the hardcover 1st edition 2001

http://www.wkap.nl/

10 9 8 7 6 5 4 3 2 1

A C.I.P. record for this book is available from the Library of Congress

Contents

Dedication

To the co-editors' families and friends for their support throughout this challenging project.

With appreciation for the loving support of our parents: Pat and Bill Chiles, Otto Erickson, and Filomena Galdikas.

To the late Antanas Galdikas, whose love of animals resonated in his children.

To Vanne Goodall, the late beloved mother of Jane Goodall.

To Lorraine P. Jenkins, the late beloved mother of Dr. Gary and Mrs. Ingrianni Shapiro. Her generosity provides for OFI's annual L.P. Jenkins Memorial Fellowship.

We intend for this volume to promote the dignity and well-being of all Africans, whether ape or human.

Acknowledgments

Many of the papers in this volume were first presented at the Third International Great Apes of the World Conference (GAWC-3) held July 3-6, 1998 in Kuching, Sarawak, Malaysia. We thank the Chief Minister of the state of Sarawak, Datuk Taib, Dr. Laila Taib, and Dr. James Masing for their support, without which GAWC-3 would not have been successful. We also thank the participants, who came from all over the world to attend the first conference on the apes held in Malaysia: Altheide, T.; Ammann, K.; Ancrenaz, M.; Andau, M.; Andayani, N.; Bates, M.; Beck, R.; Beckenbach, A.T.; Bennett, E.; Bernard, H.; Blair, S.; Blom, A.; Bloomsmith, M.; Blouch, R.; Boestani, A.N.; Bosi, E.; Boysen, S.T.; Briggs, N.; Butynski, T.; Cocks, L.; Collier, J.; Couch, R.; Czekala, N.M.; de la Rosa, M.; de Vries, H.; Didier, S.; Djojasmoro, R.; Dulek, L.; Dupain, J.; Ely, J.; Erwin, J.; Foitova, I.; Galdikas, B.M.F.; Ghaffar, N.; Gilbertson, W.; Gilmour, L.; Griffiths, M.; Hagey, L.R.; Hansen, B.; Harvey, N.C.; Hashim, N.R.; Hashimoto, C.; Hill, K.; Hof, P.; Holloway, R.; Hoole, W.S.; Howard, C.; Jamart, A.; Janecek, J.; Janke, A., Jerabkova, Z.; Jurke, M.; Jurke, S.; Kanthaswamy, S.; Karesh, W.; Kemnitz, J.; Kilbourn, A.; Komiji, J.; Lackman-Acrenaz, I.; Lardeux-Gilloux, I.; Lee, D.R.; Leiman, A.; Lowenstine, L.; MacKinnon, J.; MacKinnon, K., Martelli, P., Martens, R.; McManamon, R.; Melnick, D.J.; Minesi, C.; Mitrasetia, T.; Moore, B.; Mootnick, A.; Morales, J.C.; Mosman, K.; Muir, C.; Mutalib, A.; Nadler, R.; Nessel, J.; Nimchinsky, E.; Norcup, S.; Ollanski, M.; Onuma, M.; Paredes, G.; Patterson, F.; Perl, D.; Poullet, S.; Redmond, I.; Robertson, Y.; Rose, A.L.; Shapiro, G.; Sheeran, L.K.; Sinaga, D.W.; Singer, P.; Singleton, I.; Skrivankova, J.; Stein, T.; Stevens-Wood, B.; Stone, W.H.; Supriatna, J.; Swartz, K.B.; Tambing, E.; Tanaka, M.; Thompson, J.; Tisen, O.B.; Todd, A.; Tutin, C.; Utami, S.S.; Vancata, V.; Vancatova, M.; Van Elsacker, L.; van Hoof, J.A.R.A.M.; van Schaik, C.;

Verheyen, R.; Wisdom, L.; Wolfe, N; Young, W.; Zihlman, A.; and Zlamalova, H.

We thank the Sarawak Development Institute (SDI), particularly Lelia Sim Ah Hua, Zabariah Matali, Tang Hung Huong, and Daniel Chew, for co-organizing GAWC-3 with the Orangutan Foundation International. The SDI staff was professional, competent, and efficient during the planning stages of and at the conference itself. We also thank the staff and management of the Crowne Plaza Riverside Hotel, particularly General Manager Mohd. Khalid Buang, for hosting the event and providing excellent service to all concerned.

We thank the following agencies and organizations for sponsoring GAWC-3: Ministry of Tourism, Sarawak; Sarawak Tourism Board; Malaysia Airlines System; Malaysia Tourism Promotion Board; Orangutan Foundation International; Orangutan Foundation UK; Safeskin Corp. SDN BHD; Datuk Amar (Dr.) Leonard Linggi Jugah; Eric Raymond Foundation; Mr. Ralph Arbus; Sarawak Shell BHD; Sedgewick County Zoo; Zoo Atlanta; and Nature's Hammock.

The following agencies and individuals also provided valuable support in making the conference possible: National Park and Wildlife Department Sarawak; Police Department Sarawak; Immigration Department Sarawak; Chief Minister's Department; Kuching City North Council; Aloysious Dris; Mohd. Tuah Jais; Robert Basiuk; Sharifa Danial; K.H. Chea; Tan Hyn Teong; Ahmad Fuaad Dahlan; Ismail Hj.Talib; Abu Omar; Tony Champion; Rusty Staff; Marilyn Staff; Karin Harvey; Simone Archambeau; Michael Hawkins; and Philip Yong. Thanks also goes to the following OFI members who contributed to the conference's success: Rodney Briggs, Inggriani Hartanto Shapiro, Ashley Leiman, Pamela Breitbeil, Gloria Anderson, Filomena Galdikas, Aldona Galdikas, John Pearson, Tony Rose, Lisa Mather, Wendy Hoole, Dandy O'Shea, Chantel Bryan, and Dave Duncan.

We are grateful for the institutional support provided by Simon Fraser University (BMFG), California State University-Fullerton (LKS), and California State University-Long Beach (NB) and for our departmental colleagues' encouragement. Dr. Douglas Eernisse (Department of Biology, California State University-Fullerton) provided helpful commentary on chapter two.

We appreciate the patience and assistance of the Kluwer staff, particularly Andrea Macaluso, Deborah Doherty, Susanne St. Claire, and Mary McCarra. We express our gratitude to Marie Marley for her tireless work formatting chapters, arranging layouts, and entering our editorial changes.

Preface

Welcome to the study of great and small apes. I am delighted to discuss the great apes of Africa, this time with my colleagues Drs. Biruté Galdikas, Nancy Briggs, Lori Sheeran, and Gary Shapiro. It has been our privilege to work with manuscripts written by researchers of the highest rank. Without continuing research on the behavior of these remarkable great and small apes, the world would not understand their intelligence and complexity. This book, dedicated to the great apes of Africa, is the first in a two-volume series. The second volume focuses on the great and small apes of Asia.

I have dedicated most of my life to the great apes, specifically chimpanzees, as have my co-editors. Dr. Biruté Galdikas, my "Leakey sister" and professor at Simon Fraser University in Canada, has studied wild orangutans with unparalleled success. Dr. Nancy Briggs, professor at California State University-Long Beach, has studied great ape communication for decades. Dr. Lori Sheeran, professor at California State University-Fullerton, has studied gibbons in China and Thailand. Dr. Gary Shapiro has had the distinction of pioneering the teaching of ASL to ex-captive orangutans in the wilds of Borneo.

This first volume has been long in coming and is much needed. The world's premier primatologists, ethologists, and anthropologists present the most recent research on both captive and free-ranging African great apes. These scientists, through deep personal commitment and sacrifice, have expanded our knowledge of chimpanzees, bonobos, and gorillas.

How can we appreciate these great and small apes? By studying them and knowing their behavior and ecology, we begin to understand their great intelligence. There is a profound wisdom for all involved, including the readers of this volume, to be gained from our daily encounters with the apes, great and small. With the forests disappearing, perhaps many of these studies will never be duplicated. This is the sad dilemma all researchers face. Therefore, current events make this volume all the more valuable.

I hope you enjoy this remarkable adventure into the research on the endangered apes of Africa wherever they are found. May this volume broaden and deepen your understanding and continue to transform your relationships with these extraordinary creatures.

Jane Goodall

Volume Overview

This conference is a direct outgrowth of the first Great Apes Conference held in 1974 in Austria, where great ape experts gathered for more than one week to discuss ape issues. Conservation was absent from the agenda, and not a single paper on conservation appeared in the volume that resulted from the conference. In December 1991, for the first time scholars gathered in an ape range country, Indonesia, to discuss great apes. This time, both conservation and research were emphasized, but most discussions were about orangutans and orangutan conservation issues. African great apes were represented by the world's leading chimpanzee experts (Drs. Nishida, Goodall, Teleki, Wrangham, McGrew, and others). Only one bonobo researcher was present, and these apes, still known as "pygmy chimpanzees" to many in the audience, were scarcely mentioned.

When the third ape conference was initially planned in 1996, Indonesia was seen as the center of emerging Southeast Asian "tigers," was politically stable, and environmental issues, while severe, did not seem insurmountable. The region was suffused with the glow of economic potential, and there was talk of Indonesia becoming a regional superpower. Various economic and political events in 1997, however, exposed the underlying weakness and distortion in this picture. As toxic haze from massive forest fires in Borneo and Sumatra rose over Southeast Asia, as transportation and communication networks collapsed due to the smoke, as schools closed, and as thousands of people suffered from respiratory disorders, it became clear that Indonesia, and Southeast Asia as a whole, still had a long way to go. The abrupt economic collapse of the Southeast Asian "tigers" followed by massive devaluations of the currencies of the region occurred just as the fires were at their worst. The rupiah, long stable, plummeted from 2,500 rupiahs to the dollar to 17,000. Although the currency later regained some of its original value, the savings of millions of Indonesians were decimated and their economic futures were destroyed. At one point, every company listed on the Jakarta stock exchange was technically bankrupt. Peregrine, the large maverick brokerage firm in Hong Kong, abruptly went bankrupt when the hundreds of millions of dollars that it had invested in President Suharto's daughter's taxicab firm vanished, seemingly in the smoke shrouding the region.

In the beginning of 1998, protests swelled throughout Indonesia, reaching a crescendo of demonstrations in Jakarta during the month of May. President Suharto abruptly stepped down after several days of violent rioting in Jakarta that left 150 people dead and more than 1,000 Chinese-Indonesian women and girls raped and mutilated.

The Orangutan Foundation International decided, in light of these problems, to postpone the great apes conference until 1998. Both U.S. and British governments had issued travel advisories for their citizens visiting Indonesia, and the international media descended on the capital so that television screens around the world were filled with images of burning buildings and rioters in Jakarta. Many conference invitees, especially those with no experience in Southeast Asia, decided not to attend the conference, citing reasons of personal safety and unease even though the conference was going to be held in Malaysian, not Indonesian, Borneo.

Fresh perspectives were presented at the conference despite the loss of some invitees. An example of this was the conference's emphasis on bonobos and the strong presence of bonobo researchers and conservationists. These people came direct from the field, and they had an urgent message that had not been articulated at previous conferences. This message was that the bonobo, in terms of the species' situation in the wild, is still relatively unknown and is in grave danger of going extinct in the near future. This message had been rarely heard at conferences where wild chimpanzees and gorillas were the paramount species of interest.

Conference delegates agreed that a major strength of the conference was its focus on *all* ape species, great and small, and their conservation imperatives. The genuine hospitality of the Sarawak government and people formed the backdrop for the emerging concern for the survival of our ape relatives. Datuk Taib, the Chief Minister of Sarawak, and many of his cabinet hosted the evening reception that opened the conference, while the Minister of Tourism spoke at the plenary session that officially opened the conference's academic sessions. Equally important, those who braved the volatile situation in Southeast Asia were passionate about apes and their future, and their concern and commitment to apes were quite apparent.

Another noticeable element of this conference was the diversity of delegates and the international flavor of all the proceedings. Scholars, conservationists, students, and officials traveled from at least 20 countries. Datuk Taib used the occasion of the conference to announce the establishment of a new reserve for orangutans (Ulu Sabayu) and, visibly moved by the presentation of the Orangutan Foundation International's Lorraine P. Jenkins Fellowship for Orangutan and Rainforest Research to three graduate students (including one from Sarawak), himself announced the creation of the Sarawak Orangutan Fellowship Awards.

The presentation of the papers and formal and informal discussion culminated in the development of the Apes Declaration and a letter to the World Bank signed by numerous participants. In the spirit of the conference, the Chief Minister announced that an anti-bushmeat law would become effective immediately. This law allowed subsistence hunting of wild, unprotected species such as bearded pigs but totally banned the commercial trade, sale, and barter of such forest meat.

Not surprisingly given that the conference was held in Southeast Asia, many of the papers were on the two Asian apes, the sole surviving great ape species in Asia, the orangutan, and the nine to eleven (depending on the taxonomy used) small ape species. There simply were too many papers at this conference for one volume. In preparing the book for publication, we decided to separate the papers into the African and Asian apes, as this resulted in two volumes of approximately equal length.

The African volume includes several papers presented at the conference plus four from invited scholars. The chapters in Volume I reflect the strengths of the conference, including field research on bonobos, the great ape species arguably most threatened in the wild, the discussion of the ghastly bushmeat problem faced by African apes, and the diversity of fascinating studies, both in captive and wild situations, that reveal the high cognitive abilities of apes and the complexity of their behavior. One section of Volume I is devoted to the physiological bases for behavior and aging among great apes and further highlights ape and human similarities. Two chapters discuss models of hominid divergence from their ape cousins; both lead to fascinating, and perhaps controversial, conclusions. Finally, Peter Singer, the philosopher who gave us *Animal Liberation*, and Paolo Cavalieri, who forever changed the nature of the debate concerning animal rights, conclude Volume I with a summary of their arguments concerning the personhood of apes.

Some hopeful conservation trends are noted by several authors in Volume I, including the establishment of new reserves, new field sites, and renewed governmental interest in conservation in African apes' range countries. In the face of increasing evidence of the behavioral, cognitive, and cultural complexities of these and other apes and the unresolved debates surrounding ape and human origins and evolution, strong philosophical, anthropological, and ecological arguments can and must be made for global efforts to conserve all apes. We intend for these two volumes to further such efforts.

Biruté M.F. Galdikas, Nancy Briggs, Lori K. Sheeran, Gary L. Shapiro, and Jane Goodall

SECTION ONE
ISSUES IN APE AND HUMAN EVOLUTION

INTRODUCTION

Volume One opens with three important and compelling contributions to our understanding of ape and human evolution. As the primate group most closely related to humans, apes hold special significance for tracing the course of human evolution and teasing out which features are uniquely human versus which are shared in common with other primates or other mammals. Despite the fact that apes have long been the focus of primatologists' attention, these three chapters show that there is still much to learn, and even such fundamental data as the timing of chimpanzee and human evolutionary divergence may require further analysis.

Richard Wrangham and David Pilbeam open this section with a discussion of how living African apes are best used to model the behaviors and anatomy of the common ancestor of apes and humans. They note that the common ancestor likely shared a number of features, such as aspects of dentition and locomotion, with modern gorillas (*Gorilla*), chimpanzees (*Pan troglodytes*), and bonobos (*P. paniscus*). Wrangham and Pilbeam argue that a reduction in scramble competition in proto-bonobo populations, however, selected for reduced sexual dimorphism and relaxed aggression among bonobo males. This is reflected in neoteny in bonobo cranial development, which is considered a derived suite of characteristics in the bonobo line. They argue, conversely, that chimpanzees, and possibly early humans, evolved under conditions of more intense scramble competition, in which male competition and aggression remained comparatively high. They suggest that the use of African apes and other species are important in our attempts to model the behavior and biology of the common ancestor, rather than advocating the preferential use of one *Pan* species to the exclusion of the other.

Paleoanthropology has been revolutionized by the use of changes in genetic material to date divergence times of various taxa. In some cases, molecular evolution can be used as a "clock" to calibrate the evolution of a group, but debate centers on how reliable the "clock" is, the extent to which molecular changes correspond to phenotypic changes, and how to interpret changes in coding versus noncoding and nuclear versus mitochondrial DNA. Meanwhile, the technology used to assess molecular changes has itself evolved and has been applied to a wide variety of taxa, including many mammals.

One widely quoted date of importance in chimpanzee and human evolution is the 5 million-year divergence of these two lines (Sarich and Wilson, 1967). Molecular analysis showed that apes and humans are much more closely related than phylogenies based on anatomical changes suggested. While controversial for some years, the 5 million-year chimpanzee/human divergence date has been widely accepted by primatologists and anthropologists and has been used in part to calibrate other evolutionary events in the primate order. Recent comparisons of chimpanzee and human DNA show that the two species differ by approximately 1%, which seems to underscore the recency of the two taxa's separation. Such data have become integral to conservationists' arguments that chimpanzees and other apes share a great deal in common with humans and that they should be granted special status—perhaps even special rights (Singer and Cavalieri, this volume).

Axel Janke and Ulfur Arnason, molecular biologists with extensive experience in mammalian evolution, argue in chapter two that it is time for a critical and rigorous re-evaluation of the "5 MYBP doctrine". Their analysis yields a 10 to 13 million-year chimpanzee/human divergence date that they argue is more consistent with the paleontological records and molecular evidence of primates and other mammals. Their figure can be used to recalibrate other evolutionary events in the primate order, including ape speciation and human migrations, in a way that they believe is more consistent with all available evidence. No doubt, this might be a controversial paper, but we believe that it is important that this re-evaluation be aired.

Carol MacLeod, Karl Zilles, Axel Schleicher, and Kathleen Gibson close this section with a groundbreaking analysis of the structure of the primate brain and the implications of structural changes for ape and human cognition. Many scientists have attempted to assess primate cognition by measuring the subjects' performance on various tests. For example, the capacity for self-awareness has traditionally been tested using mirror experiments (Gallup, 1970; Heyes, 1998). Comparisons of apes' performance on these and other tests indicated that gorillas, chimpanzees, orangutans (*Pongo*), and bonobos share many features of intelligent behavior in common with humans, but that the small apes (Hylobatidae) apparently lack such capacities. Research on the aging process of ape brains likewise indicates that gibbons may lack particular neuron types that are present in the brains of great apes and humans (Erwin, *et al.*, this volume). Thus, there may be an evolutionary gulf in the cognitive capacities of great apes and humans versus all other primates.

The analyses of MacLeod and her colleagues demonstrate that gibbons, great apes, and humans are united in the large size of the lateral cerebellum (neocerebellum). This resulted in an expanded "capacity for complex movement", a distinctive characteristic of ape locomotion. In understanding the unique nature of human intelligence and cognition, paleoanthropologists must seek, therefore, the evolutionary factors that selected for neocerebellar changes in the common ancestor of the hominoids. MacLeod and her colleagues suggest that the answer may lie in locomotion and the "greater scope and complexity when feeding" afforded by the newly expanded neocerebellum. In his review of the various forms of primate intelligence, Byrne (2001:171) likewise hypothesizes that the "...need to procure food efficiently was crucial in cognitive change..." that occurred during the course of ape evolution.

MacLeod and her colleagues suggest that gibbons, infrequently the subjects of cognition tests because of their high strung and uncooperative nature, may possess some of the same "proto-human" behaviors (such as language and tool use) documented in large-bodied apes (McGrew, 1996). However, they note that the great apes and humans share a marked increase in the overall quantity and organization of other regions of the brain, which may account for further complexity in the cognitive skills of these species.

The three chapters in Section One demonstrate that even in taxa as extensively known as the apes, many intriguing questions remain. The answers hold great promise for our understanding of the human position in the natural world and further underscore the importance of protecting these, our closest living relatives, from extinction.

REFERENCES

Byrne, R.W., 2001, Social and technical forms of primate intelligence. Pp. 147-172 in: (Ed. F.B.M de Waal), *Tree of Origin: What Primate Behavior Can Tell Us about Human Social Evolution*, Cambridge, MA: Harvard University Press.

Erwin, J., *et al.*, this volume.

Gallup, Jr., G.G., 1970, Chimpanzees: Self recognition. *Science* 167: 86-87.

Heyes, C.M., 1998, Theory of mind in nonhuman primates. *Behav. Brain Sci.* 21: 101-114.

McGrew, W.C., 1996, *Chimpanzee Material Culture: Implications for Human Evolution*. New York: Cambridge University Press.

Sarich, V.M. and Wilson, A.C., 1967, Immunological time scale for human evolution. *Science* 179: 1200-1203.

Singer, P. and Cavalieri, P., this volume.

Chapter 1

AFRICAN APES AS TIME MACHINES

R. Wrangham and D. Pilbeam
Department of Anthropology, Peabody Museum, Harvard University, Cambridge, MA 02138

1.   INTRODUCTION

"About 5 million years ago forest-ranging, knuckle-walking apes—very much like living chimpanzees—evolved...into the earliest humans..." (Zihlman, 1978:4). This view has successfully challenged alternatives such as the prebrachiationist model (descent from a generalized terrestrial quadrupedal ape), the gibbon model (descent from a terrestrial gibbon), and the Miocene fossil model (descent from a thick-enameled megadont hominoid) (Latimer, *et al.,* 1981; Pilbeam, 1996).   Increasingly strong support has come from our growing confidence in the molecular evidence that human and chimpanzee lineages diverged after the split with gorillas (Ruvolo, 1997); the recognition that *Pan* is little changed phenotypically from the African ape ancestor (Groves, 1986, 1988); and the discovery that the earliest known australopithecine fossils (probably within 1-2 million years of their likely split from the chimpanzee lineage) have more chimpanzee-like features than do later species (Richmond and Strait, 2000; White, *et al.,* 1994; Wood, 1994a; Zihlman, 1996a).   For such reasons, "the common ancestor of humans and chimpanzees was probably chimpanzee-like, a knuckle-walker with small thin-enameled cheek teeth" (Pilbeam, 1996:155).

"Chimpanzee-like", in the above conclusion, refers equally to chimpanzees (*Pan troglodytes*) and bonobos (*P. paniscus*). In this chapter we ask whether we can discriminate more finely.   To what extent can the common ancestor be reconstructed as having similar features to those found currently in either *P. troglodytes* (hereafter called "chimpanzees") or *P. paniscus* (hereafter "bonobos")?

Before bonobos were well known, chimpanzees were assumed to provide a good model of the ancestral state. Then, following an early 'scala naturae' argument by Coolidge (1933), Zihlman and her colleagues proposed bonobos to be more similar to the common ancestor, on the basis that they were "the most generalized of the African apes and have many 'primitive' features, particularly the shorter humerus relative to femur" (Zihlman and Cramer, 1978:92; see also Zihlman, 1979; Zihlman, *et al.*, 1978). Although Zihlman's idea provoked considerable research into the comparative morphology of African hominoids, it has been neither fully supported nor fully refuted. The consensus is much as Wood (1994b:31) concluded, "it is at present unclear with which of the two extant species of *Pan* the modern *H. sapiens* should be compared".

In this chapter, we suggest a different kind of conclusion. We do not ask which species provides a better model. Instead, given that we know the pattern of genetic relationships, we aim to characterize traits that differ between chimpanzees or bonobos as being either homologies or homoplasies with respect to gorillas (or humans). In many cases, there is no clear answer. However, following Shea (1983) and others, we note that the pattern of ontogenetic development of chimpanzees appears to be homologous with the pattern in gorillas, whereas it is derived in bonobos. If this conclusion is correct, the pattern of cranial ontogeny in the common ancestor was chimpanzee-like, and not bonobo-like. Inasmuch as morphologies can be interpreted correctly as reflections of behavior, therefore, this suggests that the common ancestor was likely to have been more chimpanzee-like than bonobo-like in aspects of its behavior that were correlated with cranial development. Accordingly, we briefly consider the implications for behavioral evolution.

2.   DIFFICULTIES OF CHARACTERIZING THE COMMON ANCESTOR

Genetic data are sometimes invoked to aid in characterizing the common ancestor. For example, de Waal (1998:407) characterized bonobos and chimpanzees as "equally close, and equally relevant to an understanding of human evolution", and was accordingly puzzled about "why attempts are still being made to push it (the bonobo) to the sidelines" in discussions of human behavioral evolution (cf. Zihlman, 1996b). But this puzzle is easily solved. Genetic relationship is not the relevant dimension of comparison, because the rate of morphological evolution can vary from stasis across millions of years to rapid change in a few thousand years, and is therefore not necessarily correlated with the molecular clock. This means that in the absence of other kinds of information genetic relationships cannot be used to reconstruct the nature of ancestors. The relevant process is change in phenotypes, not genotypes.

The morphological similarities among chimpanzees, bonobos, and gorillas are profound, including dietary traits such as thin-enameled molars, and locomotor adaptations for arm-hanging and knuckle-walking. When size is taken into account, the similarities become even greater. With a few exceptions (such as the small incisors of gorillas), these apes are to a considerable extent "size variants in a single morphotypic series, going stepwise from the smaller *P. paniscus* to the large male gorilla" (Zihlman, *et al.*, 1978:744; see also Groves, 1970; Hartwig-Scherer, 1993; Jungers and Susman, 1984; McHenry, 1984; Shea, 1981, 1986; Taylor, 1997). Indeed, the crania of large chimpanzees are so difficult to distinguish from those of small gorillas that they were once thought to belong to their own intermediate-sized species (the kooloo-kamba, Shea, 1984a). These clearly homologous similarities explain why, until genetic data showed that humans had evolved subsequent to the split from gorillas, the three African apes were widely considered to be both each other's closest relatives, to the exclusion of humans, and even congeneric (in the genus *Pan*, Groves, 1970; Tuttle, 1968). Such fundamental similarity means that in many cases of traits differing between chimpanzee and bonobo, gorillas are sufficiently similar to these apes to allow a meaningful comparison. Extensive change in the lineage leading to humans (Hartwig-Scherer, 1993), by contrast, means that comparisons between bonobos or chimpanzees and humans offer few convincing cases of homology.

In theory, therefore, degrees of phenotypic difference from gorillas (or occasionally humans) allow us to evaluate whether traits that differ between bonobos and chimpanzees are likely to be primitive or derived. This is true whenever gorillas (or humans) are similar to only one of these other apes. Thus, if an identifiably homologous trait is shared between gorillas and only one of the *Pan* species, it is expected to have occurred in the common ancestor. The alternative version of the trait, found in the other *Pan*, would be considered derived, and absent in the ancestor.

Unfortunately, this formula is generally ineffective despite the presence of many phenotypic differences between bonobos and chimpanzees (Groves, 1986; Izor, *et al.*, 1981; Johnson, 1981; Jungers and Susman, 1984; Kinzey, 1984; Leigh and Shea, 1996; Socha, 1984; Stanyon, *et al.*, 1986). First, it is sometimes difficult to decide whether traits represent homology or homoplasy, even among the closely related set of bonobos, chimpanzees, and gorillas. For example, de Waal and Lanting (1996) note that bonobos and gorillas have wide nostrils, unlike chimpanzees. Because the genetics and functional significance of nostril width are unknown, however, the similarity of this trait in bonobos and gorillas may well be due to homoplasy. Given this uncertainty, the primitive condition cannot be confidently reconstructed. A closely related problem is the difficulty of identifying a biologically meaningful trait. Bonobos and gorillas have darker faces than chimpanzees, for instance (Groves, 1986). This suggests that dark faces are primitive, and would therefore have occurred in the ancestor. But bonobos, unlike chimpanzees or gorillas, have pink lips. This means that the character under selection might be the whole face color ("pink lips against a dark face"), rather than "dark face". If so, gorillas are similar to neither bonobos nor chimpanzees, and again, the ancestral state cannot be reconstructed. The absence of rules for selecting phenotypic, as opposed to genetic, traits means that, in theory, different researchers can reach different conclusions from the same data.

And they do. For example, the combination of short arms and long legs in bonobos was proposed by Zihlman (1979) to be homologous with australopithecines and therefore to be primitive, compared to the longer arms and shorter legs of chimpanzees. By contrast, Hartwig-Scherer (1993) noted that larger apes have longer arms in relation to leg length. She concluded that the arm:leg length ratio of bonobos was much as expected from their body length, and was accordingly homologous with the ratio in chimpanzees and gorillas. Shea (1983), on the other hand, considered the relatively short arms of bonobos to be a pedomorphic character, derived from the chimpanzee-gorilla pattern. This unresolved problem illustrates the difficulty of identifying homologous and biologically relevant traits. As a result, most morphologists agree that it is impossible to characterize the common ancestor as being, overall, more like chimpanzees or bonobos. A typical comment is McHenry's evaluation of the ancestor's postcranial anatomy: "no living species can be considered as being the least derived in all of its anatomy" (McHenry, 1984:222). Similarly Shea (1989:94) concluded "it is far from clear that *P. paniscus* is more derived in its morphology and behavior than *P. troglodytes*."

3.   DEVELOPMENTAL PATTERN: A PRIMITIVE, HOMOLOGOUS TRAIT IN CHIMPANZEES AND GORILLAS

In spite of the problems reviewed above, extensive agreement has emerged concerning the polarity of change for one particular set of traits.

"(The bonobo's) reduced masticatory apparatus and paedomorphic skull are probably not primitive, however, but instead are derived from a more robust ancestor" (McHenry, 1984:222). This conclusion has a long history (e.g. Coolidge, 1933; Tuttle, 1975), and has been extensively buttressed by comparisons of skull size (Cramer, 1977; Shea, 1983, 1984b, 1989), skull gracility (Latimer, *et al.*, 1981), face shape (McHenry, 1984), the pattern of basicranial flexion (Laitman and Heimbuch, 1984), mandible and tooth size (Cramer, 1977; Latimer, *et al.*, 1981; Zihlman and Cramer, 1978), and the degree of sexual dimorphism in teeth (Kinzey, 1984), brains, and crania (Cramer, 1977) (see also McHenry, 1984; McHenry and Corruccini, 1981). It is based on three strong allometric patterns (Shea, 1983, 1989). At any given body length, the post-cranial linear dimensions of all three African apes are strikingly close. In a similar way, skull length predicts the cranial dimensions of each of the African apes as well as if the three apes were members of a single species. Both post-cranially and cranially, therefore, the three species fall on the same ontogenetic regressions. However, when the sizes of the cranium and the post-cranial skeleton are compared to each other, only chimpanzees and gorillas fall on the same line. Bonobos are outliers because their skulls are distinctly smaller in relation to body size (Hartwig-Scherer, 1993; Shea, 1983, 1984b, 1989).

Shea (1983, 1989) has argued explicitly that the small skull of bonobos is neotenous, i.e. achieved by a change in developmental timing. This proposal is supported not only by the relatively small size of the bonobo skull, but also by its juvenilized shape compared to chimpanzees, as well as by the marked reduction in cranial and dental sexual dimorphism. A variety of other bonobo traits are also pedomorphic, such as their high-pitched calls (Groves, 1986), white anal tail tuft (Groves, 1986), and shape and pattern of sexual swelling (Dahl, 1986). Evidence of cranial neoteny suggests that selection on a few genes controlling development could be responsible for a series of changes characterized by the cranium ceasing its development relatively earlier in bonobos (i.e. in relation to body length) than in the other apes. Accordingly, it implies that the small, juvenilized skull and associated traits constitute a biologically meaningful character or set of characters, derived from an ancestral state that continues to be expressed in a homologous way in chimpanzees and gorillas.

Compared to chimpanzees and gorillas, neither orangutans (*Pongo pygmaeus*) nor australopithecines show evidence of small skulls in relation to body length. Orangutans fall on the same ontogenetic curve of brain volume to body mass as chimpanzees and gorillas (Schultz, 1941), while australopithecines are estimated to have had marginally larger brains in relation to body mass than chimpanzees (Kappelman, 1996). The conclusion that the small, juvenilized skull is a derived trait in bonobos can therefore be rejected only by arguing that the pattern of cranial ontogeny found in the other great apes has evolved independently several times. Because this is unlikely, the chimpanzee-gorilla pattern of ontogeny, including their relatively large crania in relation to body size, is appropriately regarded as a primitive character that would have been found in the common ancestor. Fossils will eventually test this prediction.

4.   THE COMMON ANCESTOR OF THE LIVING AFRICAN APES

With the exception of gorillas and humans, female great apes show a remarkably consistent pattern of body mass. Among chimpanzees, bonobos, *Australopithecus*, *Paranthropus,* and orangutans, female body mass falls in the range of 29-46 kg (Table 1). Gorilla females average at least 25 kg higher, at 71-97 kg. The elevated body mass of gorillas (two to three times the other apes) suggests a shift to a new digestive strategy, because it is their large body that supposedly gives this species its unique ability to survive and grow on foliage foods of low nutritional quality. The argument from parsimony suggests that the common ancestor of African apes, before the line leading to gorillas split off around 8-9 mybp, was more generalized, that is, still a committed ripe-fruit-eater. Accordingly, the common ancestor should have had a female size range that conforms to the other great apes, i.e., 29-46 kg.

In this size range, our 6 mybp ancestor can be expected to have shown traits shared by bonobos, chimpanzees, and gorillas. It would have been thin-enameled and knuckle-walking, and females would have had black body coats. In addition, it would have had the cranial morphology of chimpanzees rather than bonobos and, like chimpanzees, would have been committed to fruit-eating even during periods when fruits were scarce. In view of the pedomorphic character of the white anal tail tuft, high-pitched calls, and female genitals of bonobos, it seems likely that the ancestor would have been more chimpanzee-like than bonobo-like in these traits also.

However, our confidence in this resemblance between chimpanzees and the common ancestor stops there. Because of the problems with identifying homologous traits, we have less confidence in our ability to reconstruct traits such as the ratio of arm length to leg length, pelvis structure, foot morphology, face color, ear size, nostril width, or other aspects of external appearance.

Table 1. Body mass in great apes

Species		Female mass (kg)	Male mass (kg)	Reference
Orangutan	*Pongo pygmaeus abelli*	36	78	Smith and Jungers, 1997
Orangutan	*P. p. pygmaeus*	36	79	Smith and Jungers, 1997
Bonobo	*Pan paniscus*	33	45	Smith and Jungers, 1997
Chimpanzee	*Pan troglodytes verus*	42	46	Smith and Jungers, 1997
Chimpanzee	*P.t. troglodytes*	46	60	Smith and Jungers, 1997
Chimpanzee	*P.t. schweinfurthii*	34	43	Smith and Jungers, 1997
Australopithecus	*A. afarensis*	29	45	McHenry, 1992
Australopithecus	*A. africanus*	30	41	McHenry, 1992
Australopithecus	*A. robustus*	32	40	McHenry, 1992
Australopithecus	*A. boisei*	34	49	McHenry, 1992
Homo	*H. habilis*	32	52	McHenry, 1992
Gorilla	*Gorilla gorilla gorilla*	71	170	Smith and Jungers, 1997
Gorilla	*G. g. beringei*	97	162	Smith and Jungers, 1997
Gorilla	*G. g. graueri*	71	175	Smith and Jungers, 1997
Human	Central African Republic	42	48	Smith and Jungers, 1997
Human	Guatemala	46	54	Smith and Jungers, 1997
Human	Melanesia	50	58	Smith and Jungers, 1997
Human	Australia	54	60	Smith and Jungers, 1997
Human	Saudi Arabia	56	63	Smith and Jungers, 1997
Human	Japan	52	65	Smith and Jungers, 1997
Human	Denmark	62	72	Smith and Jungers, 1997
Human	Western Samoa	73	78	Smith and Jungers, 1997
Human	Median human (from above)	53	61.5	

5.   WHY ARE BONOBOS JUVENILIZED?

The principal explanation for bonobos having smaller heads, reduced sexual dimorphism, lighter bodies, and more juvenilized features than chimpanzees and gorillas is Shea's proposal that these changes resulted from selection for reduced sexual dimorphism in morphology and behavior (Shea, 1983, 1984b, 1989). Here we elaborate Shea's idea with the specific suggestion that reduced sexual dimorphism functioned to reduce aggressive behavior by adult males.

Reduced aggression, in turn, has been attributed ultimately to bonobos having larger foraging parties than do chimpanzees (Blount, 1990; Wrangham, 1986, 1993, 2000). In the most specific formulation of this idea,

Wrangham (1993) suggested that selection for the bonobo phenotype was initiated by the loss of gorillas within the distributional range of a chimpanzee-like, proto-bonobo population around 2.5 mybp. The absence of gorillas made high-quality foliage more available for proto-bonobos than for chimpanzees. As a result, proto-bonobos experienced a reduced intensity of scramble competition compared to chimpanzees (Wrangham, 2000). Reduced scramble competition allowed more stable parties, which then made several forms of aggression more dangerous and costly, and less beneficial, to the aggressors. This change in the economics of violence led through various social consequences to female-female alliances, concealed ovulation, and reduced individual vulnerability to gang attacks. All these favored a reduction in the propensity for male aggressiveness (Wrangham and Peterson, 1996).

The idea that selection for reduced male aggressiveness produces bonobo-like features is supported by data on domesticated mammals. In dogs (*Canis familiaris*), a reduction in aggressiveness compared to their wolf ancestors (*C. lupus*) is correlated with a juvenilized morphology as adults. Bonobo-like morphology in dogs includes relatively smaller heads, smaller teeth, and smaller brains than in wolves of the same body mass (Coppinger and Schneider, 1995). As in bonobos compared to chimpanzees, the reduction in cranial dimensions from wolves to dogs is around 15-20%. In a further 14 species of domesticated mammals reviewed by Hemmer (1990), domesticates also have small brains compared to their wild ancestors. We suggest that this brain reduction may have resulted from selection for reduced aggression, as we also propose for bonobos.

Although the genetic processes underlying neoteny are not well understood, it is known that selection for tameness can produce change in various correlated features (Coppinger and Schneider, 1995). For example, by selecting wild foxes (*Vulpes vulpes*) for tameness over 20 generations, Belyaev (1979) produced various dog-like (juvenilized) traits including drooping ears and curly tail. By analogy with dogs, therefore, the tameness hypothesis suggests that smaller heads, reduced sexual dimorphism, and more juvenilized features in bonobos could all result from selection for reduced male aggression.

6.   IMPLICATIONS FOR BEHAVIOR

Humans share important aspects of behavior differentially with chimpanzees and bonobos (Table 2), differences that may tend to reflect different historical pathways. Thus, if bonobo cranial neoteny indeed results from a reduced level of scramble competition, as suggested, the chimpanzee pattern of relatively intense scramble competition is reconstructed as ancestral. Accordingly, behaviors found in chimpanzees that are consequences of intense scramble competition are good candidates for ancestral phenotypes, whereas behaviors found in bonobos that are consequences of relaxed scramble competition are likely to be derived.

Table 2. Human behaviors found in chimpanzees or bonobos, but not both. "Similar to H?" asks whether chimpanzees or bonobos are more similar to humans with respect to that trait. C=chimpanzee, B=bonobo.

Behavior	Reference	Similar to H?
Lethal raiding	Wrangham, 1999	C
Traditions of material culture	Whiten, *et al.*, 1999	C
Group hunting	Stanford, 1998	C
Intense male-bonding	de Waal, 1982	C
Male dominance over females	Smuts and Smuts, 1993	C
Extensive non-conceptive sexuality	Blount, 1990	B
Friendships among adult females	Kano, 1992	B
Relatively egalitarian males	de Waal and Lanting, 1996	B
Sexual conciliatory behavior	de Waal, 1990	B
Potentially relaxed intergroup interactions	Kano, 1992	B

Note: This list is illustrative only; it is not meant to be comprehensive.

This distinction has limited value, however, because phylogenetic continuity is impossible to confirm when it must traverse the great unknowns of 5 million years of hominid evolution. And more importantly, it has no explanatory value. The reasons why a behavior is shared must still be articulated for each species, whether in terms of functions or constraints.

In the case of lethal raiding, for example, an adaptive hypothesis suggests that in chimpanzees and humans it can be ultimately attributed partly to imbalances of power that arise from scramble competition varying in its intensity between neighboring communities (Wrangham, 1999). An accompanying argument suggests that the reduction of group hunting in bonobos is an incidental outcome of psychological changes resulting from selection against intra-specific violence, rather than against group hunting *per se* (Wrangham, 1999). Even these examples remain controversial, and in other cases (such as cultural traditions, shared between chimpanzees and humans) no detailed hypotheses have been advanced.

Homoplasies may be even harder to explain but are potentially fruitful in suggesting heuristic hypotheses. For example, in the case of increased female sexuality, the ultimate cause of the homoplasy may have been a reduced intensity of scramble competition (Blount, 1990). This can be expected to have occurred in bonobos around 2.5 mybp, when individuals were first able to form relatively permanent and stable defensive alliances within groups (Wrangham and Peterson, 1996). A parallel argument might therefore suggest that female sexuality became intensified in the human lineage at a time of reduced scramble competition. The obvious time for this was when foraging strategies changed from ape-like to human-like, presumably around 1.9 mybp (Wrangham, *et al.*, 1999). Thus chimpanzees and bonobos both provide useful models in which to generate explanatory hypotheses, regardless of how we reconstruct their ancestral behaviors.

Unfortunately, there has been a tendency to polarize between favoring chimpanzee or bonobo models when reconstructing human evolution. But suppose, against all odds, that convincing evidence emerges to show that lethal raiding is 6 million years old in humans, whereas concealed ovulation is "only" 1.9 million years old. Should this matter to our sense of ourselves, that violence is 4.1 million years older than peace? Not at all. Our convergences with either species are as evolutionarily real and as behaviorally significant as our behaviors shared by common descent. They demand explanation in terms of evolutionary benefits and constraints, and both provide numerous opportunities for helping us to think about evolutionary processes in fossil hominids.

ACKNOWLEDGMENTS

Our thanks to Brian Hare for drawing our attention to the literature on dogs, to Christopher Boehm, Laurie Godfrey, Jamie Jones, Cheryl Knott, Andy Marshall, Martin Muller, Brian Shea, Andrea Taylor, and Frans de Waal for comments, and to the Institute for Human Origins, University of Arizona, for their February 2000 symposium "First Cousins: Chimpanzees and Human Origins" which provoked this paper.

REFERENCES

Belyaev, D.K., 1979, Destabilizing selection as a factor in domestication, *J. of Heredity* 70: 301-308.

Blount, B.G., 1990, Issues in bonobo (*Pan paniscus*) sexual behavior, *Amer. Anthro.* 92: 702-714.

Coolidge, H.J., 1933, *Pan paniscus*: Pygmy chimpanzee from south of the Congo River, *Amer. J. of Phys. Anthro.* 18: 1-57.

Coppinger, R. and Schneider R., 1995, Evolution of working dogs. Pp. 21-50 in: (Ed. J. Serpell), *The Domestic Dog: Its Evolution, Behaviour, and Interactions with People*, Cambridge: Cambridge University Press.

Cramer, D.L., 1977, Craniofacial morphology of *Pan paniscus*: A morphometric and evolutionary appraisal, *Contributions to Primatol.* 10: 1-64.

Dahl, J., 1986, Cyclic perineal swelling during the intermenstrual intervals of captive female pygmy chimpanzees (*Pan paniscus*), *J. of Hum. Evol.* 15: 369-385.

Groves, C.P., 1970, *Gorillas*, London: Arthur Baker.

Groves, C.P., 1986, Systematics of the great apes. Pp. 187-217 in: *Comparative Primate Biology, Vol. 1: Systematics, Evolution and Anatomy*, New York: Alan R. Liss.

Groves, C.P., 1988, The evolutionary ecology of the Hominoidea, *Annuario de Psicología* 39: 87-98.

Hartwig-Scherer, S., 1993, *Allometry in Hominoids: A Comparative Study of Skeletal Growth Trends.* Ph.D. dissertation, Zurich University, Switzerland.

Hemmer, H., 1990, *Domestication: The Decline of Environmental Appreciation.* Cambridge: Cambridge University Press.

Izor, R.J., Walchuk, S.L., and Wilkins, L., 1981, Anatomy and systematic significance of the penis of the pygmy chimpanzee, *Pan paniscus, Folia Primatologica* 35: 218-224.

Johnson, S.C., 1981, Bonobos: Generalized hominid prototypes or specialized insular dwarfs? *Current Anthro.* 22: 363-375.

Jungers, W.L. and Susman, R.L., 1984, Body size and skeletal allometry in African apes. Pp. 131-178 in: (Ed. R.L. Susman), *The Pygmy Chimpanzee*, New York, Plenum Press.

Kano, T., 1992, *The Last Ape: Pygmy Chimpanzee Behavior and Ecology*, Stanford, CA: Stanford University Press.

Kappelman, J., 1996, The evolution of body mass and relative brain size in fossil hominids, *J. of Hum. Evol.* 30: 243-276.

Kinzey, W.G., 1984, The dentition of the pygmy chimpanzee, *Pan paniscus.* Pp. 65-88 in: (Ed. R.L. Susman), *The Pygmy Chimpanzee*, New York: Plenum Press.

Laitman, J.T. and Heimbuch, R.C., 1984, A measure of basicranial flexion in *Pan paniscus*, the pygmy chimpanzee. Pp. 49-64 in: (Ed. R.L. Susman), *The Pygmy Chimpanzee*, New York: Plenum Press.

Latimer, B.M., White, T.D., Kimbel, W.H., and Johanson, D.C., 1981, The pygmy chimpanzee is not a living missing link in human evolution, *J. of Hum. Evol.* 10: 475-488.

Leigh, S.R. and Shea, B.T., 1996, Ontogeny of body size variation in African apes, *Amer. J. of Phys. Anthro.* 99: 43-65.

McHenry, H.M., 1984, The common ancestor: A study of the postcranium of *Pan paniscus, Australopithecus* and other hominoids. Pp. 201-232 in: (Ed. R.L. Susman), *The Pygmy Chimpanzee.* New York: Plenum Press.

McHenry, H.M., 1992, How big were early hominids?, *Evolutionary Anthro.* 1: 15-20.

McHenry, H.M. and Corruccini, R.S., 1981, *Pan paniscus* and human evolution, *Amer. J. of Phys. Anthro.* 54: 355-367.

Pilbeam, D.R., 1996, Genetic and morphological records of the hominoidea and hominid origins: A synthesis, *Molecular Phylogenetics and Evolution* 5: 155-168.

Richmond, B.G. and Strait, D.G., 2000, Evidence that humans evolved from a knuckle-walking ancestor, *Nature* 404: 382-385.

Ruvolo, M., 1997, Molecular phylogeny of the hominoids: Inferences from multiple independent DNA sequence data sets, *Molecular Biology and Evolution* 14: 248-265.

Schultz, A.H., 1941, Relative size of the cranial capacity in primates, *Amer. J. of Phys. Anthro.* 28: 273-287.

Shea, B.T., 1981, Relative growth of the limbs and trunk of the African apes, *Amer. J. of Phys. Anthro.* 56: 179-202.

Shea, B.T., 1983, Paedomorphosis and neoteny in the pygmy chimpanzee, *Science* 222: 521-522.

Shea, B.T., 1984a, Between the gorilla and the chimpanzee: A history of debate concerning the existence of the *kooloo-kamba* or gorilla-like chimpanzee, *J. of Ethnobiology* 4: 1-13.

Shea, B.T., 1984b, An allometric perspective on the morphological and evolutionary relationships between pygmy (*Pan paniscus*) and common (*Pan troglodytes*) chimpanzees. Pp. 89-130 in: (Ed. R.L. Susman), *The Pygmy Chimpanzee*, New York: Plenum Press.

Shea, B.T., 1986, Scapula form and locomotion in chimpanzee evolution, *Amer. J. of Phys. Anthro.* 70: 475-488.

Shea, B.T., 1989, Heterochrony in human evolution: The case for neoteny reconsidered, *Yearbook of Phys. Anthro.* 32: 69-104.

Smith, R.J. and Jungers, W.L., 1997, Body mass in comparative primatology, *J. of Hum. Evol.* 32: 523-559.

Smuts, B.B. and Smuts, R.W., 1993, Male aggression and sexual coercion of females in nonhuman primates and other mammals: Evidence and theoretical implications, *Advances in the Study of Behavior* 22: 1-63.

Socha, W.W., 1984, Blood groups of pygmy and common chimpanzees: A comparative study. Pp. 13-42 in: (Ed. R.L. Susman), *The Pygmy Chimpanzee*, New York: Plenum Press.

Stanford, C.B., 1998, The social behavior of chimpanzees and bonobos: Empirical evidence and shifting assumptions, *Current Anthro.* 39: 399-407.

Stanyon, R., Chiarelli, B., Gottlieb, K., and Patton, W.H., 1986, The phylogenetic and taxonomic status of *Pan paniscus*: A chromosomal perspective, *Amer. J. of Phys. Anthro.* 69: 489-498.

Taylor, A.B., 1997, Scapula form and biomechanics in gorillas, *J. of Hum. Evol.* 33: 529-553.

Tuttle, R.S., 1968, Quantitative and functional studies on the hands of the Anthropoidea. I: The Hominoidea, *J. of Morphology* 128: 309-364.

Tuttle, R.S., 1975, Parallellism, brachiation, and hominoid phylogeny. Pp. 447-480 in: (Eds. W.P. Luckett and F.S. Szalay), *Phylogeny of the Primates*, New York: Plenum Press.

de Waal, F.B.M., 1982, *Chimpanzee Politics: Power and Sex Among Apes*, New York: Harper and Row.

de Waal, F.B.M., 1990, Sociosexual behavior used for tension regulation in all age and sex combinations among bonobos. Pp. 378-393 in: (Ed. T. Feierman), *Pedophilia: Biosocial Dimensions*, New York: Springer.

de Waal, F.B.M., 1998, "Comment" on Stanford (1998), *Current Anthro.* 39: 407-408.

de Waal, F.B.M. and Lanting, F., 1996, *Bonobo: The Forgotten Ape*, Berkeley, CA: University of California Press.

White, T.D., Suwa, G., Asfaw, B., 1994, *Australopithecus ramidus*, a new species of early hominid from Aramis, Ethiopia, *Nature* 371: 306-312.

Whiten, A., Goodall, J., McGrew, W.C., Nishida, T., Reynolds, V., Sugiyama, Y., Tutin, C.E.G., Wrangham, R.W., and Boesch, C., 1999, Chimpanzee cultures, *Nature* 399: 682-685.

Wood, B., 1994a, The oldest hominid yet, *Nature* 371: 280-281.

Wood, B., 1994b, The age of australopithecines, *Nature* 372: 31-32.

Wrangham, R.W., 1986, Ecology and social evolution in two species of chimpanzees. Pp. 352-378 in: (Eds. D.I. Rubenstein and R.W. Wrangham), *Ecology and Social Evolution: Birds and Mammals*, Princeton: Princeton University Press.

Wrangham, R.W., 1993, The evolution of sexuality in chimpanzees and bonobos, *Human Nature* 4: 47-79.

Wrangham, R.W., 1999, The evolution of coalitionary killing, *Yearbook of Physical Anthro.* 42: 1-30.

Wrangham, R.W., 2000, Why are male chimpanzees more gregarious than mothers? A scramble competition hypothesis. Pp. 248-258 in: (Ed. P. Kappeler), *Male Primates*, Cambridge: Cambridge University Press.

Wrangham, R.W. and Peterson, D., 1996, *Demonic Males: Apes and the Origins of Violence*, Boston: Houghton Mifflin.

Wrangham, R.W., Jones, J.H., Laden, G., Pilbeam, D., and Conklin-Brittain, N.L., 1999, The raw and the stolen: Cooling and the ecology of human origins, *Current Anthro.* 40: 567-594.

Zihlman, A.L., 1978, Women and evolution, Part II: Subsistence and social organization among early hominids, *Signs: Journal of Women in Culture and Society* 4: 4-20.

Zihlman, A.L., 1979, Pygmy chimpanzee morphology and the interpretation of early hominids, *S. African J. of Sci.* 75: 165-168.

Zihlman, A.L., 1996a, Reconstructions reconsidered: Chimpanzee models and human evolution. Pp. 293-304 in: (Eds. W.C. McGrew, L.F. Marchant, and T. Nishida), *Great Ape Societies*, New York: Cambridge University Press.

Zihlman, A.L., 1996b, Looking back in anger, *Nature* 384: 35-36.

Zihlman, A.L. and Cramer, D.L., 1978, Skeletal differences between pygmy (*Pan paniscus*) and common chimpanzees (*Pan troglodytes*), *Folia Primatologica* 29: 86-94.

Zihlman, A.L., Cronin, J.E., Cramer, D.L., and Sarich, V.M., 1978, Pygmy chimpanzee as a possible prototype for the common ancestor of humans, chimpanzees, and gorillas, *Nature* 275: 744-746.

Chapter 2

PRIMATE DIVERGENCE TIMES

A. Janke and U. Arnason
*Division of Evolutionary Molecular Systematics, Department of Genetics, University of Lund,
Sölvegatan 29, S-223 62 Lund, Sweden*

1.   INTRODUCTION

In recent years molecular data and phylogenetic analyses have been used to fill various phylogenetic gaps and to give insight into evolutionary processes.   Changes in DNA sequences, such as base substitutions, accumulate progressively, and thus DNA sequences continually record the past. Molecular sequences also make it possible to analyze a much larger amount of data than does any morphological approach. In principle, the entire genome, some 3 billion nucleotides, can be used to infer relationships and divergence times. Currently, however, the largest data sets of an ever-growing number of mammalian representatives are completely sequenced mitochondrial genomes, which allow analysis of some 10,000 nucleotides of homologous protein coding genes. Molecular studies on divergence times rely on outside information, such as that of paleontology, to calibrate the "molecular clock" (Wilson, *et al.*, 1987).

The paleontological record, especially that of primates, contains many gaps (Martin, 1990, 1993) and allows reconstruction of divergence times for very few lineages. Before a divergence becomes detectable on the basis of morphological differentiation, an unknown amount of time must pass (Wrangham and Pilbeam, this volume). For this reason, analysis of the paleontological record can give only minimum estimates for the origin of a group. Despite the specific problems with the primate fossil record (Martin, 1990, 1993), it has been used as local reference points to calibrate molecular evolutionary rates and subsequently to date primate divergence times using molecular data. In particular, the divergence of *Homo* and chimpanzee

(Sarich and Wilson, 1967) has attracted much attention, and this dating in turn has been used to date migration events of modern human ethnic groups. A rigorous investigation of such an outstanding topic as primate divergences has been neglected in recent years, despite inconsistencies of the initial approach.

2.   ESTIMATES OF CHIMPANZEE AND HUMAN DIVERGENCE TIMES

2.1   The 5 MYBP Doctrine

Human evolution has been the focus of interest since the beginning of molecular phylogenetics.   The divergences among the Hominidae were among the first to be estimated and were dated on the basis of molecular information (Sarich and Wilson, 1967).   The authors assumed that the divergence of Cercopithecoidea and Hominoidea had taken place 30 MYBP (million years before present) and used that paleontological reference point to conclude that *Homo* and chimpanzee had diverged 5 MYBP.   Similar dates have been calculated by assuming a divergence of Old and New World monkeys at 35 to 40 MYBP (Goodman, 1996; Goodman, *et al.*, 1987) and a divergence of Cercopithecoidea and Hominoidea at 25 MYBP (Porter, *et al.*, 1997).   A split of orangutans from the Hominidae at 14 to 17 MYBP has also been assumed (Adachi and Hasegawa, 1995; Sibley and Ahlquist, 1987).

For the last 30 years it has been common practice to conform various dating estimates to the originally proposed divergence time of *Homo* and chimpanzee at 5 MYBP.   This date has become a doctrine despite conspicuous inconsistencies of this approach when divergence times or estimations of evolutionary rates based on primate calibration points were applied to taxa outside the primate lineage.   For example, assuming a divergence of Old and New World monkeys at 35 to 40 MYBP results in an origin of eutherians (placental mammals) some 65 to 70 MYBP, which is entirely untenable in the light of eutherian paleontology (Archibald, 1996; Rich, *et al.*, 1997).   As a second example, some authors set the split of the orangutan from the Hominidae at 14 to 17 MYBP and date the Hominidae and chimpanzee split at ≈4 MYBP (Adachi and Hasegawa, 1995; Easteal and Herbert, 1997).   This is younger than the oldest australopithecine fossils and implies a secondary loss of bipedalism in *Pan*!

2.2   Evidence Contradicting the Doctrine

Paleontologists who voiced opposition but who were unfamiliar with molecular phylogenetic approaches were soon silenced and accepted this 4-5 MYBP doctrine.   It has been suggested that the 14 MYBP fossil *Rampithecus*, originally interpreted as a hominoid (Simons and Pilbeam, 1965), must be outside the hominoid group, since *Rampithecus* fossils are older than the hominoid origin as postulated by the Sarich-Wilson estimate. Moreover, based on a *Homo* and chimpanzee divergence of 5 MYBP, new and contradicting fossil evidence such as the >9 MYBP fossil of *Ouranopithecus*, which apparently is part of the *Austrolopithecus/Homo* lineage (de Bonis, *et al.*, 1990; de Bonis and Koufos, 1997), has been largely ignored.   Other recent paleontological finds, such as Asian anthropoids, also contradict the Sarich-Wilson estimate, as they are older than is accepted for the hominoid group (Jaeger, *et al.*, 1998).

3.   NEW APPROACHES TO PRIMATE DIVERGENCE TIMES

3.1   Use of Non-Primate References

Recently an entirely different approach has been applied to calibrate molecular rates and to date primate divergences.   Realizing the inconsistencies of the former approaches, Arnason and colleagues (1996, 1998) applied two non-primate references that are paleontologically well established and, because of their definition within narrow time limits, superior to any primate fossils for calibrating the evolutionary rates of molecular data.   The separation of Artiodactyla (Ruminantia) and the cetaceans is marked by a significant morphological change reflecting cetacean adaptation to aquatic life.   Combined analysis of molecular and paleontological data of cetaceans dates their divergence from the Artiodactyla to 60 MYBP (A/C-60) (Arnason and Gullberg, 1996). Applying the A/C-60 standard to eutherian, avian, and reptilian lineages yielded divergence times that are consistent with their respective fossil records and other lines of evidence (Härlid, *et al.*, 1998; Janke and Arnason, 1997; Janke, *et al.*, 1997).   The other reference used is the intra-ordinal separation of Equidae (horse and donkey) and Rhinocerotidae (rhinoceros) set at 50 MYBP (E/R-50) (Arnason, *et al.*, 1998; Xu, *et al.*, 1996), which also yields estimates that are consistent with eutherian evolution (Arnason, *et al.*, 2000).

Molecular data have in recent years significantly contributed to resolving open questions of eutherian (Arnason, *et al.*, 1997; Janke, *et al.*, 1994, 1997) and primate phylogeny (Arnason, *et al.*, 1996, 1998). The large number of characters, their unambiguous definition, and the stochastic nature of nucleotide substitution make the analysis of molecular data more favorable than morphological data. Morphological data permit *only* a cladistic approach, taking into account shared derived characters on the basis of maximum parsimony (MP). Molecular data, because of the stochastic nature of substitutions, enable analysis applying distance methods, e.g. neighbor joining (NJ) (Saitou and Nei, 1987) or the maximum likelihood method (ML) (Felsenstein, 1981). Because of potential pitfalls of any analytical procedure, different analytical approaches should be applied, and they should result in a consistent phylogeny. Resampling methods such as bootstrapping (Felsenstein, 1985) and statistical methods such as the Wilcoxon Rank Test (Templeton, 1983) or differences in likelihoods and their standard error (Kishino and Hasegawa, 1989) can be applied to molecular data to determine the support for individual branches and topologies. Only after a phylogeny has been well established can one attempt to estimate divergence times.

3.2   Reassessment of Primate Divergence Times

Figure 1 shows one ML tree and primate divergence times based on amino acid sequences from 11 mitochondrial protein coding genes. All branches are well supported by different methods of phylogenetic reconstruction. The branch lengths represent the genetic distances as substitutions per site. These branch lengths can be used to estimate divergence times. Another approach is the use of genetic distances as obtained by a distance matrix, where the pairwise differences between all species have been calculated (Table 3). Possible homoplasies due to parallelisms or convergence resulting in too low distance values can be accounted for by an evolutionary model of substitutions (e.g.,Hasegawa, *et al.*, 1985; Jukes and Cantor, 1969). The ML tree and estimates of branch lengths are based on an *a priori* model of sequence evolution, which also takes homoplasies into account.

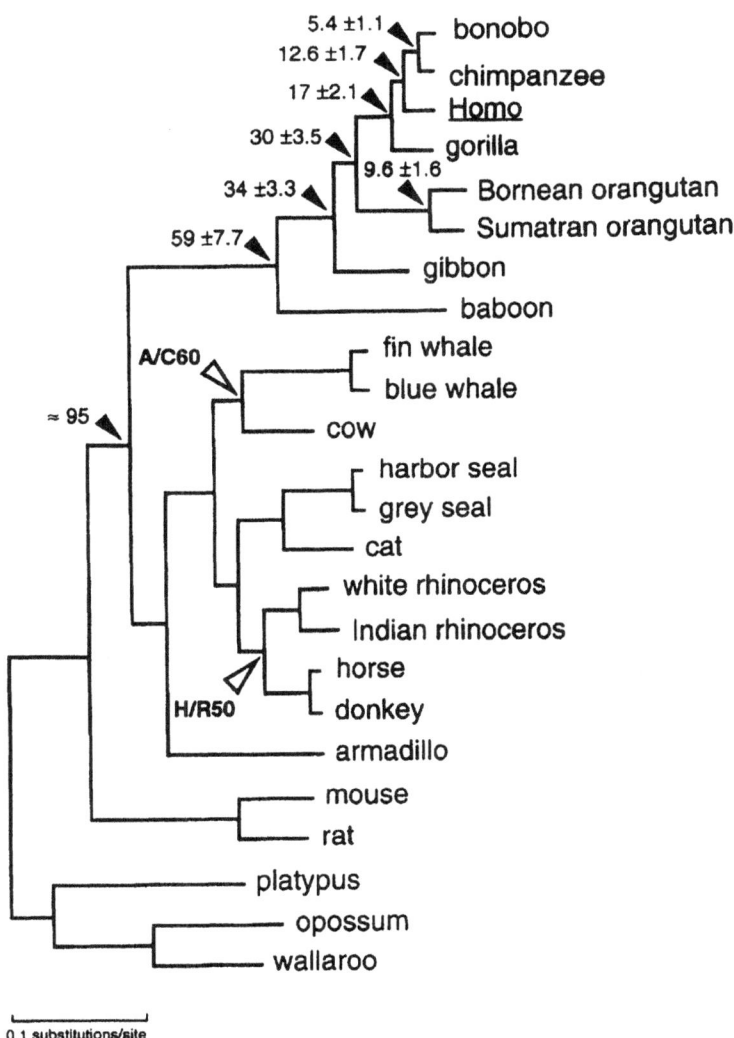

Figure 1. The primate lineage has longer branches compared to those of other eutherians.

Table 3. Paiwise differences of mammalian mitochondrial amino acid sequences.

Species	Oan	Dvi	Mro	Mmu	Rno	Dno	Pvi	Hgr	Fca	Eca	Eas	Csi
Oan	**Oan**	722	687	835	825	787	779	777	792	757	755	775
Dvi	.2780	**Dvi**	485	833	827	774	787	790	781	764	761	772
Mro	.2656	.1717	**Mro**	820	818	762	744	741	751	711	716	723
Mmu	.3375	.3289	.3252	**Mmu**	309	714	710	712	694	681	674	679
Rno	.3338	.3283	.3259	.1031	**Rno**	707	701	703	676	667	654	660
Dno	.3121	.3027	.2971	.2719	.2692	**Dno**	550	555	566	523	521	529
Pvi	.3077	.3076	.2906	.2696	.2645	.2002	**Pvi**	50	322	384	373	371
Hgr	.3068	.3092	.2888	.2708	.2658	.2023	.0157	**Hgr**	322	385	374	373
Fca	.3134	.3031	.2912	.2614	.2535	.2044	.1079	.1078	**Fca**	358	353	361
Eca	.2972	.2969	.2740	.2558	.2496	.1877	.1327	.1329	.1222	**Eca**	45	250
Eas	.2961	.2949	.2766	.2532	.2447	.1871	.1283	.1286	.1204	.0142	**Eas**	248
Csi	.3050	.2988	.2777	.2561	.2475	.1894	.1278	.1284	.1235	.0828	.0822	**Csi**
Run	.3081	.3027	.2896	.2549	.2509	.1947	.1308	.1297	.1291	.0875	.0877	.0497
Bta	.2977	.2954	.2807	.2595	.2562	.1954	.1386	.1429	.1380	.1200	.1217	.1261
Bph	.3156	.3206	.3045	.2857	.2808	.2130	.1772	.1788	.1817	.1573	.1555	.1608
Bmu	.3174	.3236	.3042	.2833	.2791	.2151	.1769	.1801	.1835	.1568	.1556	.1589
Pha	.4006	.3944	.4042	.3605	.3541	.3369	.3154	.3141	.3153	.2917	.2914	.2882
Hla	.3656	.3578	.3669	.3136	.3131	.2905	.2770	.2772	.2787	.2591	.2584	.2549
Ppy	.3844	.3751	.3775	.3312	.3282	.3061	.2929	.2929	.2985	.2801	.2791	.2671
Pab	.3811	.3724	.3715	.3357	.3345	.3070	.2932	.2942	.2971	.2777	.2776	.2704
Ggo	.3746	.3605	.3642	.3230	.3205	.2895	.2745	.2760	.2779	.2608	.2588	.2520
Ppa	.3719	.3555	.3604	.3189	.3135	.2819	.2707	.2734	.2701	.2563	.2550	.2483
Ptr	.3702	.3517	.3592	.3162	.3120	.2770	.2690	.2717	.2705	.2519	.2515	.2472
Hsa	.3765	.3567	.3625	.3163	.3141	.2832	.2717	.2736	.2735	.2570	.2553	.2486

Table 3 Continued

Species	Run	Bta	Bph	Bmu	Pha	Hla	Ppy	Pab	Ggo	Ppa	Ptr	Hsa
Oan	781	758	785	789	953	878	911	905	897	892	888	900
Dvi	779	762	810	815	946	876	908	899	882	872	864	872
Mro	748	727	770	768	958	884	904	890	884	875	872	877
Mmu	676	691	741	735	892	792	825	831	809	804	796	794
Rno	668	683	734	729	875	792	822	832	806	793	789	792
Dno	543	540	579	582	845	747	778	775	743	729	716	731
Pvi	379	401	495	494	802	718	749	749	713	708	702	706
Hgr	377	413	501	504	799	717	749	750	715	713	707	709
Fca	377	399	509	513	798	728	770	763	725	709	708	714
Eca	263	351	445	443	749	683	724	717	686	677	665	676
Eas	263	356	440	440	748	681	721	715	681	673	663	671
Csi	154	369	457	451	745	676	700	704	669	663	658	660
Run	**Run**	382	462	462	747	678	708	707	669	667	663	670
Bta	.1310	**Bta**	409	404	775	701	739	738	692	686	673	674
Bph	.1625	.1414	**Bph**	82	796	735	762	759	728	724	716	720
Bmu	.1633	.1401	.0263	**Bmu**	805	739	765	765	738	734	728	732
Pha	.2885	.3009	.3147	.3189	**Pha**	568	599	615	584	561	558	571
Hla	.2572	.2674	.2851	.2880	.2078	**Hla**	395	397	329	325	332	333
Ppy	.2717	.2853	.2988	.3012	.2208	.1385	**Ppy**	155	366	356	360	359
Pab	.2730	.2857	.2985	.3023	.2281	.1397	.0511	**Pab**	351	350	354	349
Ggo	.2531	.2636	.2823	.2879	.2141	.1130	.1264	.1215	**Ggo**	172	173	170
Ppa	.2511	.2608	.2802	.2857	.2038	.1110	.1225	.1206	.0564	**Ppa**	71	139
Ptr	.2502	.2560	.2770	.2839	.2030	.1135	.1241	.1221	.0567	.0228	**Ptr**	131
Hsa	.2539	.2561	.2790	.2856	.2090	.1144	.1243	.1209	.0559	.0453	.0427	**Hsa**

Distances were calculated from 11 concatenated mitochondrial protein coding genes (3193 aa) used by Arnason, *et al.* (1998). Above diagonal: absolute number of amino acid differences, below diagonal: inferred number of substitutions per site according to the mtREV-24 model of amino acid sequence evolution (Adachi and Hasegawa, 1996). Species are abreviated by the first letter of the genus and the first two letters of the species name (see Figure 1): Oan-platypus, Dvi-opossum, Mro-wallaroo; Mmu-mouse; Rno-rat; Dno-armadillo; Fca-cat; Pvi-harbor seal; Eca-horse; Csi-white rhinoceros; Bta-cow; Bph-fin whale; Bmu-blue whale; Pha-hamadryas baboon; Hla-gibbon; Ppy-Bornean orangutan; Pab-Sumatran orangutan; Ggo-gorilla; Ppa-bonobo; Ptr-chimpanzee; Hsa-*Homo.*

If the evolutionary rate is the same in the species used for calibration and the species for which the divergence time should be determined (they evolve at a strictly clock-like rate), then the unknown divergence time is simply a function of the time and distances or time and branch lengths of the species involved. Primates, however, have an accelerated evolutionary rate relative to other eutherians. Figure 1 shows that the primate lineage has longer branches compared to those of other eutherians. This hypothesis is also supported by a relative rate test (Sarich and Wilson, 1973) or a likelihood ratio test. Phylogenetic analysis of the mitochondrial cytochrome *b* protein coding gene indicates that the acceleration took place very early in primate evolution (Arnason, *et al.*, 1998). The acceleration may have occurred before the separation of New and Old World monkeys, and perhaps before the separation of prosimians and anthropoids (some lemurs show a faster evolutionary rate compared to other eutherians). The genetic data, although based on a limited data set, are available from a larger number of primate species and can be taken as an indication of the origin of the accelerated evolutionary rate in the primate lineage.

Since the exact onset of this acceleration is unknown, we assume that it occurred shortly after the split of the primates from other eutherians. A rate difference (ΔR) between primates and the calibration points can then be calculated from distances or as a ratio of ML branch lengths. To calculate the distance between the ancestor of primates and the reference used, an indisputable but not too distant outgroup can be used to enhance detection of rate differences. We use the myomorph rodents as such an outgroup. The references are cows and whales when utilizing the A/C-60 or horses, donkeys, and rhinoceroses when the E/R-50 reference point is applied. The ΔR is calculated as:

$$\Delta R = \frac{(D_{\text{rod-prim}} + D_{\text{prim-ref}} - D_{\text{rod-ref}})/2}{(D_{\text{rod-ref}} + D_{\text{prim-ref}} - D_{\text{rod-prim}})/2} = a/b$$

where $D_{\text{rod-prim}}$ is the distance between myomorph rodents and primates, $D_{\text{prim-ref}}$ is the distance between primates and the reference, and $D_{\text{rod-ref}}$ is the distance between myomorph rodents and the reference. Figure 2 illustrates how the distances between an outgroup (rodents), the primates, and reference are related to each other. The calculated ΔR represents a ratio of the branch leading from a hypothetical ancestor to the primates (a) and the reference (b) respectively. In this way the rate difference can be calculated independently of the fossil record. The ΔR is then used to normalize the distances between the reference and the investigated species. The primates have, depending on the evolutionary model and references used, a 1.4-1.6 times faster evolutionary rate than the references (Arnason, *et al.*, 1996, 1998). Figure 1 also shows the divergence times as the mean values of the estimates calculated by Arnason, *et al.* (1996, 1998). These estimates are reasonably consistent regardless of the dataset used (second codon position or amino acid sequences), the reference points, or the evolutionary models (rate homogeneity or heterogeneity among sites) that were applied to calculate separation times from distances or branch lengths.

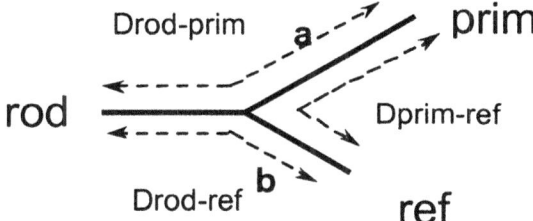

Figure 2. The distances between an outgroup (rodents), the primates, and reference.

Estimating primate divergence times based on molecular data and the non-primate references for calibrating evolutionary rates result in dates for various divergences that are about two times older then previously assumed. Thus, human and chimpanzee did not diverge 4-6 MYBP but 10-13 MYBP. Since the relative divergence times among different primate lineages are independent of the analyzed data set (mitochondrial or nuclear data) (Arnason, *et al.*, 1996), the differences can be attributed to the ages allocated to the reference points.

The three reference points that have so far been applied by other authors to date primate separations are based exclusively on divergences within the primate lineage: the divergence of Old and New World monkeys at 30-40 MYBP (Goodman, 1996; Goodman, *et al.*, 1987; Porter, *et al.*, 1997; Sarich, 1970), the cercopithecoid/hominoid split set at 25 to 30 MYBP (Porter, *et al.*, 1997; Sarich and Wilson, 1967; Sibley and Ahlquist, 1987) and the divergence between *Pongo* and *Gorilla/Pan/Homo* set at 14 to 17 MYBP (Adachi and Hasegawa, 1995; Sibley and Ahlquist, 1987). Application of any these reference points to date the separation of other primate lineages results in the same divergence time of humans and *Pan* at about 5 MYBP. Invalidating any one of these calibration points would automatically invalidate the others.

Using any of the above-mentioned primate divergences as a reference point to estimate non-primate divergence times would result in a eutherian origin of <70 MYBP, which is incompatible with eutherian paleontology (Rich, *et al.*, 1997). Likewise the Artiodactyl-Cetacea divergence would be calculated at <30 MYBP. This is an impossible estimate given that the fossil record shows differentiation into individual cetacean lineages by that time (Fordyce and Barnes, 1994). It would be possible, however, to invoke *ad hoc* hypotheses, such as a number of complex increases and decreases in molecular evolutionary rates, to reconcile these data with the 5 MYBP divergence of humans and *Pan* as established by Sarich and Wilson (1967).

3.3   Fossil Data that Challenge the Doctrine

An even more direct test of the 5 MYBP doctrine comes from recent fossil findings that were not known by the previously mentioned authors. These fossil data invalidate two of the three primate reference points previously used:

*   The assumption of a divergence of Old and New World monkeys at 30-40 MYBP is refuted, as anthropoid fossils from the late Early and Middle Eocene ($\approx$50 MYBP) have been described (Godinot and Mahboubi, 1992).
*   A human/chimpanzee split at 5 MYBP is refuted when the >9 MY old *Ouranopithecus* fossils are placed on the lineage leading to *Homo* (Begun and Kordos, 1997; de Bonis and Koufos, 1997; de Bonis, *et al.*, 1990). This fossil genus has been recently accepted as being outside the divergence between *Homo* and *Pan*, and its divergence has been set at 10 MYBP (Andrews, *et al.*, 1997). This divergence time is twice what was originally calculated by Sarich and Wilson (1967).

4.   REASSESSING THE DOCTRINE

4.1   Primate Divergence Times

Figure 3 illustrates the discrepancies between the original primate divergences compared to what is estimated in this paper. Primate datings according to the doctrine (black circles) are younger than the age of the new fossil findings (shaded bars). These estimates are directly invalidated because a differentiation into two groups in the fossil record cannot predate a molecular estimate of the same split. The entire line of reasoning following the doctrine collapses. Conversely, the calculations (Arnason, *et al.*, 1996, 1998) of primate divergence times (open circles) are not refuted by the primate fossil record, and furthermore show how the fossil record in some cases significantly underestimates the origin of a specific group. Martin (1993) dealt with the problem of underestimation of divergences in interpretations of the fossil record. Extrapolation of the assumptions made by the 5 MYBP doctrine (black circles and line) would result in an eutherian origin at $\approx$70 MYBP, an unsound proposition under the most conservative interpretation of the fossil record (Rich, *et al.*, 1997).

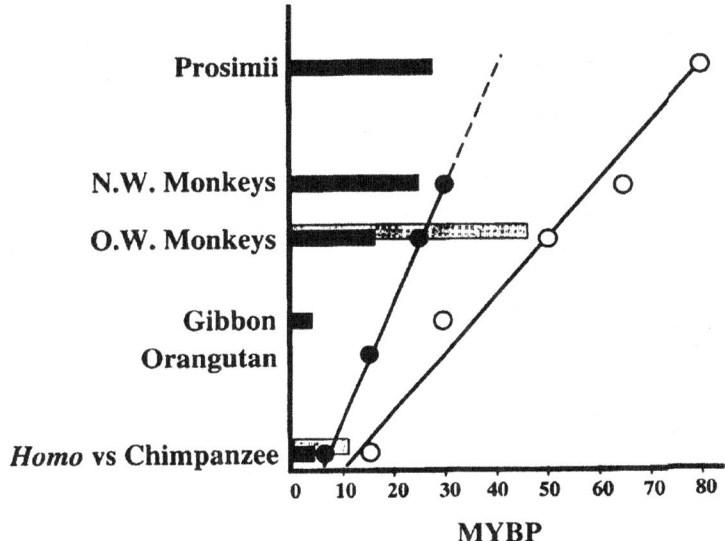

Figure 3. Discrepancies between the original primate divergences (solid circles) compared to what is estimated in this chapter (open circles).

4.2   Human Migration

Studies estimating the migration events of *Homo sapiens* rely on the separation of humans and chimpanzees at 5 MYBP as a reference date. Genetic differences of mitochondrial DNA among various human ethnic groups indicate that Australian aborigines migrated to Australia ≈40,000 years ago (Cann, *et al.*, 1987). This is in conflict with recently described artifacts that date to more than 100,000 years ago (Bahn, 1996; Fullagar, *et al.*, 1996) and are associated with *Homo sapiens*. Australia must have been populated at least 100,000 years ago, more than two times earlier than previously recognized. A colonization event 100,000 years ago is, however, consistent with a human and chimpanzee split of 10-13 MYBP.

4.3   Orangutan Divergence

Xu and Arnason (1996) recently described a pronounced genetic difference in the mitochondrial DNA sequences of Bornean and Sumatran orangutans. Despite only minor morphological differences, these two taxa show a larger genetic difference than many other closely related species. For example, the difference between the two orangutan populations is ≈80% of that between *Homo* and *Pan* (Table 4). Phylogenetic analysis indicates that, compared to other primates, the evolutionary rate in the mitochondrial genome of the orangutan is further accelerated (Arnason, *et al.*, 1996). This can also be seen in Figure 1: the branches leading to the orangutans are longer compared to those leading to other primates. Taking the rate difference into account, the two orangutan populations are genetically more different than other well recognized eutherian species (Table 4). Clearly the Sumatran and Bornean orangutan should be considered distinct species.

Table 4. Genetic difference of the Bornean and Sumatran orangutan compared to that of other closely related eutherian species pairs.

Species Pair	aa Distance	Distance %	Corrected Distance %
Bornean orangutan/ Sumatran orangutan	0.0511	100	100
Homo/Chimpanzee	0.0440	80	112
Blue whale/Fin whale	0.0263	52	94
Harbor seal/Grey seal	0.0157	31	56
Horse/Donkey	0.0142	28	51

The aa distance values are taken from Table 3. The two orangutans were set to 100% and compared to the other species pairs. The rate differences for orangutan relative to *Homo*/chimpanzee $\Delta R=1.3$ and primates (excluding orangutan) relative to other eutherians $\Delta R=1.4$ (Arnason, *et al.*, 1996) were used to calculate the corrected percent distance values.

The divergence time of Bornean and Sumatran orangutans is estimated to be 7-11 MYBP (Arnason, *et al.*, 1996). The long separation time leading to the extensive genetic difference between these two taxa should been taken into account in breeding and conservation programs. Crossbreeding should be avoided in order to maintain their unique diversity (Xu and Arnason, 1996). The genetic difference observed in mitochondrial DNA can be used as an argument to give the two orangutans seperate species status, which may in turn also support efforts to ensure their conservation.

The orangutan example demonstrates that morphological and molecular evolution can occur at extremely different paces and can thus bias the interpretation of the fossil record. Regardless of the completeness of the orangutan fossil record, the high degree of morphological similarities and a separation of the Bornean and Sumatran orangutan at 10 MYBP would not have been detected from the fossil evidence alone.

5. CONCLUSION

Reevaluation of primate divergence times was necessary after it became obvious that the originally proposed dates are in conflict with the paleontological record. The recently established non-primate calibration points (AC-60 and ER-50) provide divergence times of eutherians and primates that are consistent with the fossil record. Although the divergence times among primates proposed by Arnason, *et al.* (1996, 1998) may seem unorthodox, they are in agreement with the paleontological record and provide a framework for reevaluating primate evolution.

ACKNOWLEDGMENTS

We thank Suzette Mouchaty for valuable comments. This work was supported by The Erik Philip-Sörensens Foundation, the Swedish Natural Sciences Research Council, and an EU grant (Molecular Phylogeny of Mammalian Orders: A Model Study).

REFERENCES

Adachi, J. and Hasegawa, M., 1995, Improved dating of the human-chimpanzee separation in the mitochondrial DNA tree: Heterogeneity among amino acid sites, *J. Mol. Evol.* 40: 622-628.

Adachi, J. and Hasegawa, M., 1996, Model of amino acid substitution in proteins encoded by mitochondrial DNA, *J. Mol. Evol.* 42: 459-468.

Andrews, P., Begun, D.R., and Zylstra, M., 1997, Interrelationships between functional morphology and paleoenvironments in Miocene hominoids. Pp. 29-58 in: (Eds. Begun, D.R., Ward, C.B., and Rose, M.D.), *Function, Phylogeny and Fossils: Miocene Hominoid Evolution and Adaptation*, Plenum Press, New York.

Archibald, J.D., 1996, Fossil evidence for late Cretaceous origin of "hoofed" mammals, *Science* 272: 1150-1153.

Arnason, U. and Gullberg, A., 1996, Cytochrome b nucleotide sequences and the identification of five primary lineages of extant cetaceans, *Mol. Biol. Evol.* 13: 407-417.

Arnason, U., Gullberg, A., and Janke, A., 1997, Phylogenetic analyses of mitochondrial DNA suggest a sister group relationship between Xenarthra (Edentata) and Ferungulates, *Mol. Biol. Evol.* 14: 762-768.

Arnason, U., Gullberg, A., and Janke, A., 1998, Molecular timing of primate divergences as estimated by two non-primate calibration points, *J. Mol. Evol.* 47: 718-727.

Arnason, U., Gullberg, A., Janke, A., and Xu X., 1996, Pattern and timing of evolutionary divergences among hominoids based on analyses of complete mtDNAs, *J. Mol. Evol.* 43: 650-661.

Arnason, U., Gulberg, A., Gretarsdottir, S., Ursing, B., and Janke, A., 2000, The mitochondrial genome of the sperm whale and the establishment of a new molecular reference for estimating eutherian divergence dates, *J. Mol. Evol.* 50: 569-578.

Bahn, P.G., 1996, Futher back down under, *Nature* 383: 577-578.

Begun, D.R. and Kordos, L., 1997, Phyletic affinities and functional convergence in *Dryopithecus* and other Miocene and living hominids. Pp. 291-316 in: (Eds. Begun, D.R., Ward, C.B., Rose, M.D.), *Function, Phylogeny, and Fossils: Miocene Hominoid Evolution and Adaptatio*, New York: Plenum Press.

de Bonis, L. and Koufos, G., 1997, The phylogenetic and functional implications of *Ouranopithecus macedoniensis*. Pp. 317-326 in: (Eds. Begun, D.R., Ward, C.B., Rose, M.D.), *Function, Phylogeny, and Fossils: Miocene Hominoid Evolution and Adaptation* New York: Plenum Press.

de Bonis, L., Bouvrain, G., Geraads, D., and Koufos, G., 1990, New hominid skull material from the Miocene of Macedonia in northern Greece, *Nature* 345: 712-714.

Cann, R.L., Stoneking, M., and Wilson, A.C., 1987, Mitochondrial DNA and human evolution, *Nature* 325: 31-36.

Easteal, S. and Herbert, G., 1997, Molecular evidence from the nuclear genome for the time frame of human evolution, *J. Mol. Evol.* 44 (Supplement 1): S121-S132.

Felsenstein, J., 1981, Evolutionary trees from DNA sequences: A maximum likelihood approach, *J. Mol. Evol.* 17: 368-376.

Felsenstein, J., 1985, Confidence limits in phylogenisies: An approach using the bootstrap, *Evolution* 39: 793-791.

Fordyce, R.E. and Barnes, L.G., 1994, The evolutionary history of whales and dolphins *Annu. Rev. Earth Planet. Sci.* 22: 419-455.

Fullagar, R., Price, D.M., and Head, L.M., 1996, Early human occupation of northern Australia: Archaeology and thermoluminescence dating of Jinmium rock shelter, Northern Territory, *Antiquity* 70: 751-773.

Godinot, M. and Mahboubi, M., 1992, Earliest known simian primate found in Algeria, *Nature* 357: 324-326.

Goodman, M., 1996, Epilogue: A personal account of the origins of a new paradigm, *Molec. Phylogenetics Evol.* 5: 269-285.

Goodman, M., Miyamoto, M.M., Czelusniak, J., 1987, Pattern and process in vertebrate phylogeny revealed by coevolution of molecules and morphologies. Pp. 140-176 in: (Ed. Patterson C.), *Molecules and Morphology in Evolution: Conflict or Compromise?*, New York: Cambridge University Press.

Härlid, A., Janke, A., and Arnason, U., 1998, The complete mitochonrial genome of *Rhea americana* and early avian divergencies, *J. Mol. Evol.* 46: 669-679.

Hasegawa, M., Kishino, H., and Yano, T., 1985, Dating of the human-ape splitting by a molecular clock of mitochondrial DNA, *J. Mol. Evol.* 22: 160-174.

Jaeger, J.J., Chaimanee, Y., and Ducrocq, S., 1998, Origin and evolution of Asian hominoid primates. Paleontological data versus molecular data, *Comptes Rendus de l'Académie des Sciences*, Paris, Sciences de la Vie 321: 73-78.

Janke, A., and Arnason, U., 1997, The complete mitochondrial genome of *Alligator mississippiensis* and the separation between recent archosauria (birds and crocodiles), *Mol. Biol. Evol.* 14: 1266-1272.

Janke, A., Xu, X., and Arnason, U., 1997, The complete mitochondrial genome of the wallaroo (*Macropus robustus*) and the phylogentic relationship among Monotremata, Marsupialia and Eutheria, *Proc. Natl. Acad. Sci., USA* 94: 1276-1281.

Janke, A., Feldmaier-Fuchs, G., Thomas, W.K., von Haeseler, A., and Pääbo, S., 1994, The marsupial mitochondrial genome and the evolution of placental mammals, *Genetics* 137: 243-256.

Jukes, T.H. and Cantor, C.R., 1969, Evolution of protein molecules. Pp. 21-123 in: (Ed. Nunro, H.N.), *Mammalian Protein Metabolism*, New York: Academic Press.

Kishino, H. and Hasegawa, M., 1989, Evaluation of the maximum likelihood estimate of the evolutioary tree topologies from DNA sequence data, and the branching order in Hominoidea, *J. Mol. Evol.* 29: 170-179.

Martin, R.D., 1990, *Primate Origins and Evolution: A Phylogenetic Reconstruction*, New Jersey: Princeton University Press.

Martin, R.D., 1993, Primate origins: Plugging the gaps, *Nature* 363: 223-234.

Porter, C.A., Scott, P.L., Czelusniak, J., Schneider, H., Schneider, P.M.C., Sampaio, I., and Goodman, M., 1997, Phylogeny and evolution of selected primates as determined by sequences of the e-globin locus and 5' flanking regions, *Int. J. Primatol.* 18: 261-295.

Rich, T.H., Vickers-Rich, P., Constantine, A., Flannery, T.F., Kool, L., and van Klaveren, N., 1997, A tribosphenic mammal from the Mesozoic of Australia, *Science* 278: 1438-1442.

Saitou, N. and Nei, M., 1987, The neighbor-joining method: A new method for reconstructing phylogenetic trees, *Mol. Biol. Evol.* 4: 406-425.

Sarich, V.M., 1970, Primate systematics with special reference to Old World monkeys. Pp. 175-266 in: (Eds. Napier, J.R., Napier, P.H.), *Old World Monkeys: Evolution, Systematics and Behaviour*, New York: Academic Press.

Sarich, V.M. and Wilson, A.C., 1967, Immunological time scale for human evolution, *Science* 158: 1200-1203.

Sarich, V.M. and Wilson, A.C., 1973, Generation time and genomic evolution in primates, *Science* 179: 1144-1147.

Sibley, C.G. and Ahlquist, J.E., 1987, DNA hybridization evidence of hominoid phylogeny: Results from an expanded data set, *J. Mol. Evol.* 26: 99-121.

Simons, E.L. and Pilbeam, D., 1965, Preliminary revision of the Dryopithecinae (Pongidae, Anthropoidea), *Folia Primatol.* 3: 81-152.

Templeton, A., 1983, Phylogenetic inference from restriction endonuclease cleavage site maps with particular reference to humans and apes, *Evolution* 37: 221-244.

Wilson, A.C., Ochman, H., and Prager, E.M., 1987, Molecular time scale for evolution, *Trends Genet.* 3: 241-247.

Wrangham, R. and Pilbeam, D., this volume.

Xu, X. and Arnason, U., 1996, The mitochondrial DNA molecule of Sumatran orangutan and a molecular proposal for two (Bornean and Sumatran) species of orangutan, *J. Mol. Evol.* 43: 431- 437.

Xu, X., Janke, A. and Arnason, U., 1996, The complete mitochondrial DNA sequence of the greater Indian rhinoceros, *Rhinoceros unicornis*, and the phylogenetic relationship among Carnivora, Perissodactyla, and Artiodactyla (+ Cetacea), *Mol. Biol. Evol.* 13: 1167-1173.

Chapter 3

THE CEREBELLUM: AN ASSET TO HOMINOID COGNITION

C.E. MacLeod[1], K. Zilles[2], A. Schleicher[3], and K.R. Gibson[4]

[1]*Department of Anthropology, Langara College, 100 West 49 Avenue, Vancouver, BC, Canada V5Y 2Z6*

[2]*Director, C. & O. Vogt Institute for Brain Research, University of Duesseldorf, PO Box 10 10 07, D-40001 Duesseldorf, Germany; Director, Institute of Medicine, Forschungzentrum Juelich GmbH, D-52425 Juelich, Germany*

[3]*C. & O. Vogt Institute for Brain Research, University of Duesseldorf, PO Box 10 10 07, D-40001 Duesseldorf, Germany*

[4]*Chair, Department of Basic Sciences, Dental Branch, University of Texas, Houston, TX 77225*

1. INTRODUCTION

The human brain in its size and organization remains the single most important anatomical adaptation of the genus *Homo*, and details of its evolution are a source of intense curiosity. The human brain can be compared to the brains of our closest relatives to reveal what aspects of anatomy and organization are unique or outstandingly developed in our species. However, our brain is also a product of primate evolution as a whole and hominoid evolution in particular. To understand the brain, we must look beyond ourselves and explore events in brain evolution that are deeper than the last few million years, events that we have shared with the apes.

35

Studies of brain evolution have tended to concentrate on the most progressive structure of the brain, the neocortex. The neocortex does not exist in splendid isolation, but sends and receives information to and from the rest of the brain in a complex of interacting functional systems. One such system includes the cerebellum, which receives information from cortical and subcortical structures through the cerebellar peduncles. The cerebellum receives a massive input from the neocortex via the pontine nuclei in anthropoids and sends information back through the thalamic nuclei not just to primary motor and sensory cortex, but to widespread areas of the neocortex, including premotor, oculomotor, and prefrontal cortices (Middleton and Strick, 1994, 1997a, 1997b), temporal and posterior parietal lobes, and paralimbic cerebral cortices (Schmahmann and Pandya, 1995).

The contribution of the cerebellum to brain functioning has been largely neglected in studies of the evolution of the brain because the cerebellum has been seen as purely motor, concerned only with balance and coordination. With the advent of functional brain imaging, however, and the pioneering work of Leiner, Leiner, and Dow (1986, 1989, 1991, 1993), our view of the cerebellum is radically changing. Both experimental and clinical data show that the cerebellum, especially its lateral lobes, participates in a number of cognitive activities once thought to be confined to the neocortex (Schmahmann, 1991, 1996). These higher functions include: the planning of complex movements (Thach, 1996, 1997); sensory discrimination and the integration of sensory and motor functions at a fundamental level (Gao, *et al.*, 1996); visuo-spatial problem solving with a strong cognitive component (Kim, *et al.*, 1994); direction of visually-based attention and the ability to switch from one modality to another (Akshoomoff and Courchesne, 1992; Allen, *et al.*, 1997; Courchesne, *et al.*, 1994); procedural learning (Doyon, 1997; Molinari, *et al.*, 1997); and particular linguistic functions in humans, such as the creation of verbs from nouns (Fiez and Raichle, 1997; Fiez, *et al.*, 1992; Raichle, *et al.*, 1994). The representation of all of these higher functions is most concentrated in the lateral cerebellum, or the neocerebellum.

In gross anatomy, the cerebellum can be divided into the medial part, or the vermis, and the lateral part, or the hemispheres. In higher primates, the hemispheres are dominated by the neocerebellum, which is phylogenetically and ontogenetically the most progressive part (Voogd, *et al.*, 1990). The three zones of archicerebellum (vermis), paleocerebellum (paravermis), and neocerebellum (lateral cerebellum) have distinct functions and specific efferent nuclei (the fastigial, interposed, and dentate nuclei, respectively). Gross anatomy can be linked with function if the volume of the cerebellar hemispheres is taken to represent the neocerebellum, and the volume of the dentate nucleus its output. The principal inferior olivary nucleus is an additional source of input to the neocerebellum and is part of a tight feedback loop with the dentate nucleus (Figure 4). It is thus possible to measure a functionally integrated system and, through volumetric comparisons in ape, monkey, and human brains, to draw conclusions about the relative importance of some neurological activities.

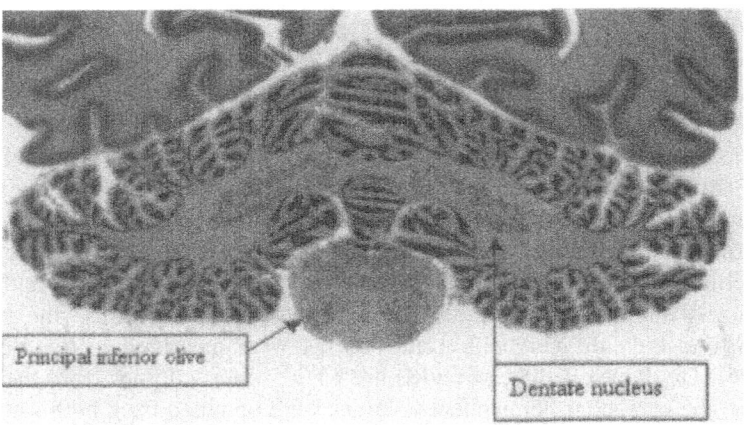

Figure 4. Coronal section of a gibbon cerebellum, with the dentate and inferior olivary nuclei apparent.

2. METHODOLOGY

This study measures brains and brain parts from two very distinct samples. One is a collection of *in vivo* anthropoids scanned with magnetic resonance (MR) imaging from the Yerkes Regional Primate Research Center in Atlanta, Georgia. Eleven species are represented, including two *Gorilla gorilla*, four *Pan paniscus*, seven *Pan troglodytes*, four *Pongo pygmaeus*, four *Hylobates lar*, six *Homo sapiens s.*, six *Macaca mulatta*, four *Cercocebus atys*, two *Papio cynocephalus*, four *Cebus apella*, and four *Saimiri sciureus*. The other sample is from a collection of *post mortem* fixed and mounted brains at the C. & O. Vogt Institut für Hirnforschung in Duesseldorf, Germany. Some of the apes from the Stephan collection are included in the Hirnforschung sample. The Hirnforschung sample encompasses 16 species of anthropoids, including three *Gorilla gorilla*, two *Pan paniscus*, seven *Pan troglodytes*, four *Pongo pygmaeus*, five *Hylobates lar*, eight *Homo sapiens s.*, one *Erythrocebus patas*, two *Macaca mulatta*, two *Cercopithecus* (sp. unknown), one *Cercocebus albigena*, two *Papio cynocephalus*, three *Ateles paniscus*, two *Alouatta* (sp. unknown), one *Cebus apella*, two *Saimiri sciureus*, and two *Aotus trivirgatus*. Each database has 47 primates, including a combined total of 42 ape brains representing all extant ape genera. This is the largest sample of ape brain and brain-part volumes ever compiled. A sample of this size permits a more rigorous statistical treatment using regression analysis because the data can be reliably subdivided into "monkey and ape" or "monkey and hominoid" groupings or grades. Grade shifts (Martin, 1980) can be tested for significance with multiple regression. Shozo Matano and colleagues measured many of the same structures in an attempt to discern cerebellar patterning in primate evolution (Matano, 1992; Matano and Hirasaki, 1997; Matano, *et al.*, 1985a, b), but the considerably larger sample size of our study shifts attention to broader tendencies in hominoids over monkeys at the level of grade, rather than species-specific distinctions.

Structures measured in the histological sections include the whole brain, cerebellum, vermis, dentate nucleus and principal inferior olivary nucleus. MR scans do not provide the resolution necessary to measure the small nuclei, and only the cerebellum and vermis were measured in the Yerkes sample. Dr. James Rilling provided the whole brain volumes from the MR scans. The cerebellar hemisphere volumes were obtained from both samples by subtracting the volume of the vermis from the cerebellum. Although different techniques were used to measure the vermis in the histological sections and MR scans, the two data sets were in remarkable accord for these and other common measurements, and thus provided a cross-validation of the accuracy of the volumes.

The MR scans comprising the Yerkes sample are T1-weighted using a gradient-echo protocol, with slice thickness at 1.2 mm. and slice interval at 0.6 mm. The resolution of each voxel is $0.588mm^3$. Before measurement with the Philips "Easy Vision" software, the images were first formatted and standardized to the Talairach plane, oriented along the anterior and posterior commissures (Talairach and Tournoux, 1988). The cerebellum and vermis were measured in the sagittal plane, and boundaries could be verified on the frontal and horizontal planes that appeared on the screen, enabling a three-dimensional view of the structure. The intensity threshold of the pixels in the scan was set interactively. The intensity threshold "saturates" the soft tissue and distinguishes it from the hard tissue or the meningeal fluid. With the vermis, the area of interest was outlined interactively with a lasso-like contour drawing procedure that permitted fine adjustments to the boundaries of the vermis. To minimize measurement error, the intensity threshold set for the measurement of the vermis was identical to that of the cerebellum. All images in the stack were measured, yielding the volume of the target structure.

The serially-sectioned brains at the Hirnforschung Institute were also digitized and measured interactively using intensity threshold, but with a program specially developed at the Institute based on the KS400 system. The histological sections themselves were digitized and stored in the computer in order to measure the volumes of whole brain, cerebellum, and vermis. The area of interest was outlined, threshold intensity set, and the area of each section measured. Volumes were calculated by multiplying the sum of the areas of each structure by the distance between sections, each structure averaging 25 sections. All fixed volumes were multiplied by a shrinkage factor that corrected for the inevitable shrinkage of the tissue during fixation, embedding, and other histological procedures. The dentate and principal inferior olive were subject to similar measurement techniques to the histological sections but were first traced from the mounted sections at a magnification of 17.5 times before being digitized into the computer. A minimum of 12 sections were traced and measured for the determination of the volume of each structure, more than is needed for accurate volume estimates (Zilles, *et al.*, 1982). The nuclei were also corrected to absolute values by the application of the same shrinkage factor used for the histological sections.

3.   RESULTS

The percentage of the brain devoted to cerebellar hemisphere is much greater in hominoids compared to monkeys (Figure 5), but ratios give an incomplete picture of hominoid hemisphere expansion. This is because brain parts increase with their own particular exponents as a function of overall brain increase in evolution, and more progressive structures with higher exponents of increase occupy increasing percentages of total brain volume as the brain expands (Finlay and Darlington, 1995). Regression analysis exposes any differential increase in the structures measured that would be beyond predicted allometry. The method of multiple regression used in this study allows the testing for significance and slope of the additional x-value of grade. In some cases, the data are better explained with two regression lines representing two grades than with one single regression line. Table 5 summarizes some of the regressions (performed with Systat 9). The reader is referred to MacLeod (2000) for details of the statistical analysis.

Table 5. This table summarizes some of the regressions (performed with Systat 9). See MacLeod (2000) for details of the statistical analysis.

Regression	r2 value	SE	Derived Formulae	Grade Sig.	Parallel
Hemisphere to Vermis	.964	.266	y' (monkeys) = 0.429 + 1.399x y' (hominoids) = 1.529 + 1.332x	yes	yes
Predicted y-values (antilogs)[1]			y (monkeys) = 8.9 cc.'s y (hominoids) = 24.6 cc.'s		

Hominoid hemispheres are 2.8 times greater than monkey hemispheres with a vermis of the same volume.

Hemispheres to Whole Brain Minus Cerebellum (No humans in regression)[2]	.983	.162	y' (monkeys) = -3.226 + 1.136x y' (apes) = -2.283 + 1.053x	yes	yes
Predicted y-values (antilogs)			y (monkeys) = 16.2 cc.'s y (apes) = 26.8 cc.'s		

Ape hemispheres are 66% larger than monkey hemispheres with a whole brain minus cerebellum of the same volume.

Principal inferior olive to Cerebellum	.933	.257	y' (monkeys) = 1.177 + .745x y' (hominoids) = 1.519 + .741x	no[3]	yes
Predicted y-values (antilogs)			y (monkeys) = 49.9 mm^3 y (hominoids) = 69.2 mm^3		

The hominoid principal inferior olive is 39% larger than a monkey's principal inferior olive in a cerebellum of the same volume.

Dentate to Hemisphere No humans	.936	.223	y' (monkey) = 2.949 + .845x y' (ape) = 2.442 + .863x	no[3]	yes
Predicted y-values (antilogs)			y (monkey) = 132.8 mm^3 y (ape) = 83.4 mm^3		

A monkey dentate is 60% larger than an ape dentate with a hemisphere of the same volume.

[1] The y-value is calculated from the median x-value. Both logged values are then returned to their natural numbers to determine differences of magnitude between two grades.

[2] Humans destroy the symmetry of the grade shift because their hemispheres are below expected, as discussed in the text. The shift is still significant with humans in the basic model.

[3] The full regression model with interaction is too sensitive to determine significant grade shifts in every case, largely because of co-linearity. However, in both the above dentate and principal inferior olive regressions, the grade shifts are significant when the two x-variables are not allowed to interact (basic model).

The dramatic increase in the hemisphere to vermis proportion in hominoids compared to monkeys is the most striking pattern in the evolution of cerebellar circuitry in the anthropoids (Figure 6). Hominoids, including gibbons and humans, have cerebellar hemispheres that are 2.8 times what one would expect from a monkey brain with a vermis of a given size. When regressed against the rest of the brain, the cerebellar hemispheres retain this grade shift between hominoids and monkeys (Figure 7). The increase in the hemisphere proportions accounts for the increase in the cerebellum size in hominoids compared to monkeys beyond predicted allometry.

The expansion of the nuclei was not in tandem. The principal inferior olivary nucleus increased with the cerebellar hemispheres and showed a positive grade shift relative to the whole brain volume (Figures 8 and 9). The dentate nucleus did not show a positive grade shift relative to the whole brain volume and actually showed a negative grade shift relative to the cerebellar hemispheres. The human dentate was far below the volume that would be expected from either monkey or ape hemisphere and cerebellar volumes (Figures 10 and 11). Although a strong correlation was seen between the expansion of the principal inferior olive and the cerebellar hemispheres, two genera, *Aotus* and *Saimiri*, were dropped from this regression because of statistical incompatibility. *Aotus* was also a strong outlier with the dentate nucleus, but not *Saimiri*. A much larger sample size would be needed before anything could be concluded, but other regressions indicate that the evolutionary pattern of the dentate and the principal inferior olive differs somewhat from the evolution of the cerebellar hemispheres.

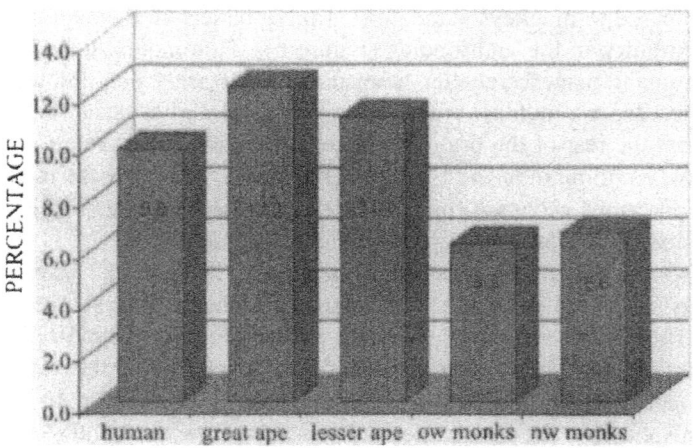

Figure 5. The percentage of cerebellar hemisphere to whole brain by grade. Hominoids are significantly different from monkeys (p<.0001).

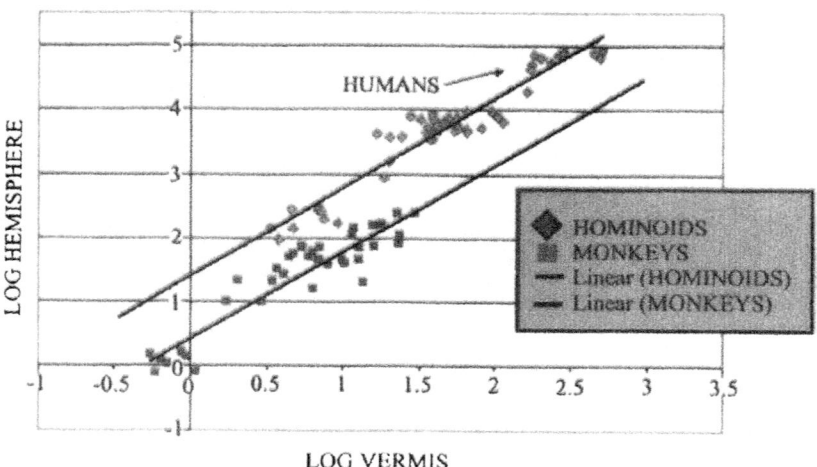

Figure 6. Logged hemisphere to logged vermis volumes from a combined sample showing hominoid and monkey grades.

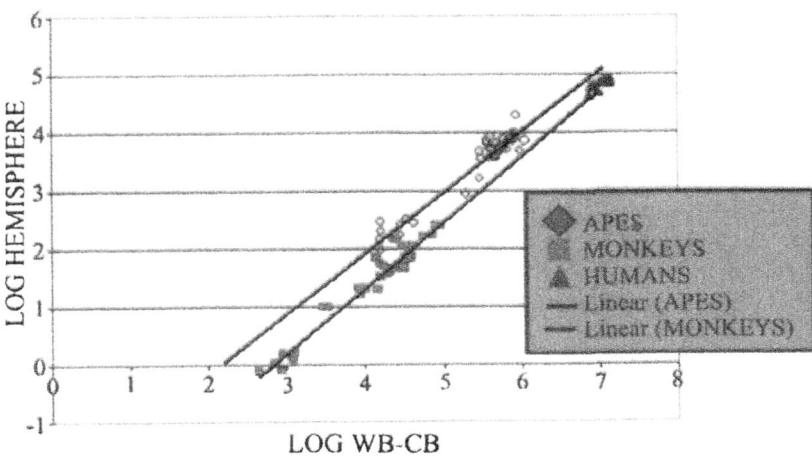

Figure 7. Logged hemisphere volume to the logged whole brain minus cerebellum volume from a combined sample.

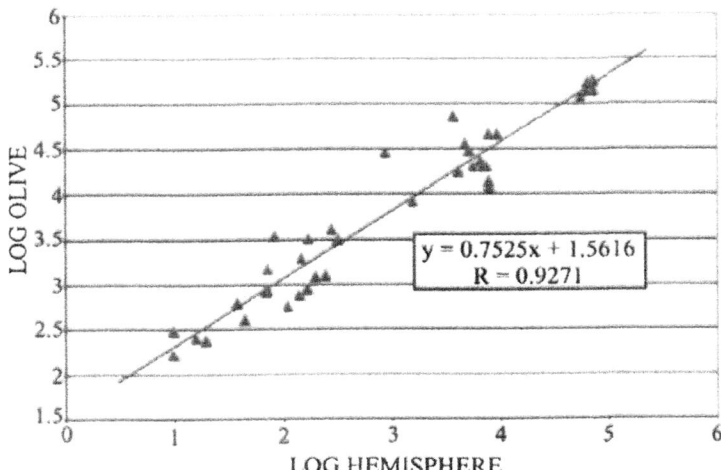

Figure 8. Logged volume of the principal inferior olivary nucleus regressed against the cerebellar hemispheres. Data are best explained by a single regression line, indicating the co-expansion of the inferior olive with the hemispheres. Note that the smallest-brained monkeys in the sample, *Aotus* and *Saimiri*, are excluded from the regression because of their extreme outlier values.

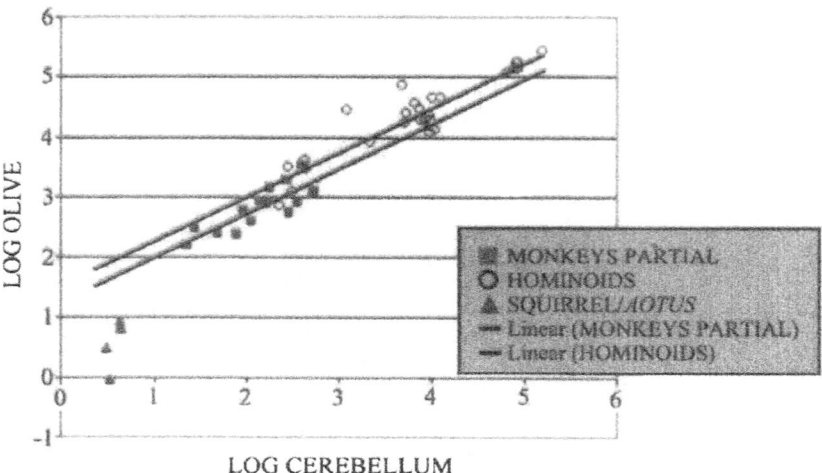

Figure 9. Logged principal inferior olive volume regressed against the logged volume of the cerebellum. *Aotus* and *Saimiri* are excluded from the regression line as outliers. The inferior olive shows a positive grade shift relative to the cerebellum because of the conservative structure of the vermis included in cerebellar volume.

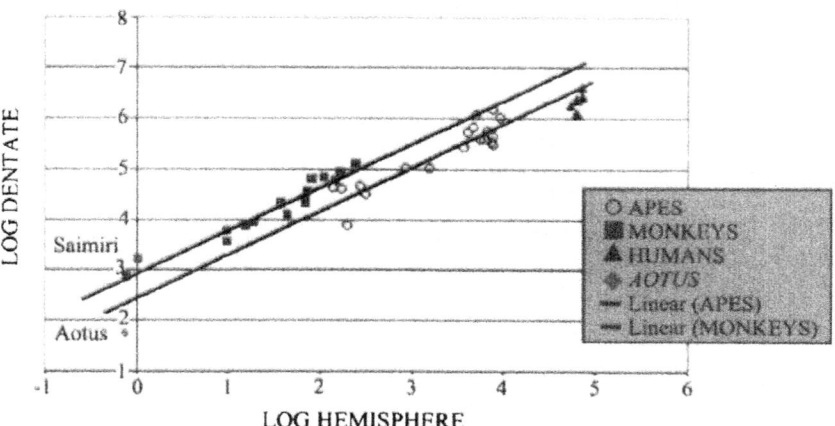

Figure 10. Logged dentate nucleus volumes against the logged cerebellar hemisphere volumes. The *Saimiri* are excluded from the regression line but fit in with the other anthropoids, whereas the single *Aotus* is too anomalous and is excluded. Notice in this regression that the hominoids are considerably below the regression line for monkeys, a reversal of the pattern of the other regressions. Human dentate volumes are below the regression line established for the apes.

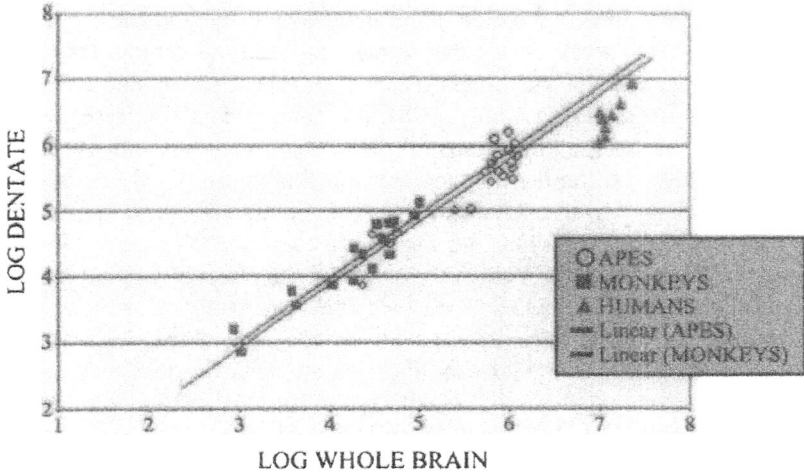

Figure 11. Logged dentate volume regressed against the logged volume of the whole brain. The two regression lines established for monkeys and apes are not significantly different but are significantly parallel. Humans are excluded from the ape regression line because their values are considerably below the rest of the hominoids.

4.   CHANGES IN CEREBELLAR PROPORTIONS

The uniformity and symmetry of the results, in which the grade shift is shown by an increase in proportion of one brain part to another but in which the exponent of increase of that brain part relative to the whole brain remains constant, supports the argument that the change in cerebellar proportions occurred in the common ancestor to the hominoids. Although there is species-specific variation within the sample, this variation is never enough to pull the grades of hominoid and monkey off of their parallel slopes in the hemisphere to vermis regression. In the hemisphere to vermis relation, both gibbons and humans fit on the regression line with the rest of the apes, and all apes and humans are clearly distinct from the monkeys.

The differential increase in the volume of the cerebellum beyond allometric constraints has been documented by Rilling and Insel for the Yerkes database (1998), but our study attributes this expansion to the increase in the lateral over the medial cerebellum, enabling a more precise functional interpretation of the data. The hemisphere to vermis expansion in the hominoids is not confined to the great apes, but clearly applies to all hominoids, including gibbons. The human cerebellum and hemispheres, when regressed against the rest of the brain, are slightly below what would be expected from the regression line determined by the rest of the hominoids. Further work on the Hirnforschung Institute sample shows a differentially expanded neocortex in humans over other anthropoids and thus confirms Rilling and Insel's view (1998) that a differential increase in the size of the neocortex in humans increases whole brain volume disproportionately, pulling the human cerebellar volumes to below expected values (i.e., a change in the denominator). The complete set of regressions clearly show that humans share the increase in the cerebellar hemispheres and the reorganization of cerebellar circuitry with the rest of the hominoids.

The pattern of nuclear increase is somewhat surprising. Leiner, Leiner and Dow (1986, 1989, 1991, 1993) and Matano and colleagues (1985b; Matano and Hirasaki, 1997) are of the opinion that the dentate nucleus expanded far beyond what would be expected in humans. This view should be revised in the light of the Hirnforschung data, which represent a five-fold increase of ape dentate nuclei volumes. The dentate is the efferent nucleus to the neocerebellum and appears to be expanding only with the general increase in the whole brain volume (Figure 11). In humans especially, it is not a progressive structure. By contrast, the principal inferior olive increases in proportion to the jump in cerebellar hemisphere size. It sends information specifically to the neocerebellum and the dentate nucleus and contributes to the increase in the information processing capacity of the cerebellum as shown by the larger lateral cerebellar cortex. The inferior olive affects the functioning of the Purkinje cells, the "workhorses" of the cerebellum that send the inhibitory signals to the deep cerebellar nuclei. The climbing fibers of the inferior olive embrace the Purkinje cells in synaptic contacts and cause them to fire. They play a crucial role in the timing and patterning of cerebellar activity, particularly with regard to the learning of new motor patterns (Albus, 1971; Marr, 1969). Our study demonstrates not only an increase in cerebellar hemisphere volume, but also a remodelling of cerebellar circuitry in hominoids through the mosaic expansion of essential nuclei with their particular functions.

The cerebellum receives information from both central and peripheral nervous systems, and appears to break up the sensory data into smaller parts and place these parts in different somatotopic contexts on the cerebellar cortex, in what is termed "fractured somatotopy" (Welker, 1987). Whether this reorganization of information applies to all cerebellar afferents remains to be seen (Schmahmann, 1996), but the structure of cerebellar organization would indicate that it is able to process information in a unique way and is not parroting the information processing pattern of the cerebral cortex. The neocortex and the cerebellum are engaged in a dialectic, and the increase in the size of the lateral hemispheres in hominoid evolution augmented the capacity of the brain to process sensory information from all modalities. Certainly the ability to perceive and process information from diverse sources and to organize an integrated response to this complexity is fundamental to intelligent behavior, and the increase in lateral cerebellar cortex is a crucial neurological substrate to the evolution of this behavior in the higher primates.

5.   NEUROPSYCHOLOGICAL FUNCTIONS

The medial and anterior cerebellum are the locus of the execution of motor patterns, and the lateral cerebellum is the area for the planning of these movements (Thach, 1997). The increase in the ratio of lateral to medial cerebellum would imply that the hominoids have an enhanced capacity for complexity of movement. Anyone who has seen the arc of a gibbon's swing from one branch to the next in a complicated three-dimensional pattern is impressed by its graceful dance. Even the great apes move in unusual and flexible ways when clambering in the trees, in contrast to the somewhat stereotypical movement of macaques (Povinelli and Cant, 1995). It is the cognitive aspect of movement, the wider range of choices for movement, the more complicated choreography that hanging suspended from the trees requires, that is the purview of the lateral cerebellum. If it were simply a question of better locomotor coordination then the vermis, associated with balance, equilibrium, and execution of movement, would have increased in volume at a much greater rate. Instead, the vermis is a conservative structure that shows no positive grade shift.

Only some of the early apes showed the characteristic suspensory anatomy characteristic of living hominoids, but all showed more versatile locomotor abilities closer to those of spider monkeys or chimpanzees (Fleagle, 1999). The proconsulids of the early Miocene were predominantly frugivorous (Fleagle, 1999), as were the cercopithecoids during this time (Benefit, 1999). Certainly, the expanded lateral cerebellum could have contributed to the early hominoid's success in their niches by allowing for a greater scope and complexity of movement when feeding, both over and under branches. The feeding adaptation may have been the primary reason for the postulated differential expansion of the cerebellar hemispheres in the common hominoid ancestor. Frugivory is associated with larger ranges (Clutton-Brock and Harvey, 1979; Milton, 1981), which in turn require more mapping abilities to return to fruiting trees that are patchily dispersed. This would involve the posterior part of the brain, including the occipital and parietal lobes and the cerebellar hemispheres. Of the massive number of fibers that descend to the cerebellum from the neocortex, it is estimated that

10 million of these are from the parietal and occipital lobes, including their associative areas (Stein, *et al.*, 1987). However, these afferent connections are not reciprocal, as the dentate efferents project more to the rostral brain, including frontal and prefrontal areas (Middleton and Strick 1997a, b), areas traditionally associated with "thinking", strategy, and choice. Thus, functional connectivity with a higher cerebellar component would allow for increased visuo-spatial mapping abilities and frontal lobe targeting of viable fruiting trees.

In a functional magnetic resonance imaging (fMRI) study, the dentate nucleus showed large bilateral activation during a cognitive task that required human subjects to exchange the position of four blue pegs with four red pegs on a board while following a set of rules (Kim, *et al.*, 1994). The "insanity task" was far from easy. It was contrasted with a task that was visually guided but required no cognitive component. The dentate activity in the "insanity task", three to four times greater than in the simple visually guided task, indicates that the neocerebellum contributed substantially to cognitive processing. This contribution is probably tied to another neuropsychological function of the cerebellum, that of procedural learning. In a positron emission tomography (PET)-activation study of the generation of verbs from nouns, Raichle and colleagues recorded activity in the right lateral cerebellum as well as the left frontal and anterior cingulate gyrus (Raichle, *et al.*, 1994). After ten minutes of practice, the reponses to the task became quicker, more accurate, and more stereotyped, while the PET scan showed a decrease in activity in the cerebellum and prefrontal cortex, implying that the cerebellum was more active in the *learning* of the cognitive activity than in its execution. Other experimental evidence shows greater cerebellar activation for visuo-spatial tasks and working memory tasks when the tasks are first being learned (Fiez and Raichle, 1997). These insights into neocerebellar functioning in humans could also apply to apes in the domain of feeding strategy, notably extractive foraging.

Complex, hierarchical processing in extracting food is characteristic of the great apes and humans in particular (Gibson, 1986; Parker and Gibson, 1977, 1979). In omnivorous extractive foraging, nutrients must be removed from matrices in which they are embedded or encased. The logical sequence of operations necessary to extract the foods is a cognitive precursor to tool use (Parker and Gibson, 1977, 1979). Procedural learning is important to successful extractive foraging, and by implication, to the successful use of tools demonstrated by apes in laboratory settings. The superior performance of apes over monkeys in language and tool-based tasks could be due in part to their larger cerebellar hemispheres and augmented inferior olivary learning circuit, specifically with regard to visuo-spatial tasks and procedural learning.

The cerebellum has been called a sensory-motor interface (Bower, 1977), with the cerebellar hemispheres in particular involved with the more active and reactive exploratory data acquisition involving the somatosensory and auditory systems. In an fMRI experiment, Gao and colleagues (1996) found that the dentate nucleus and the neocerebellum were most active in tasks in which subjects had to determine whether the shapes or the textures of two objects matched. Although the null hypothesis of the experiment predicted the greatest involvement for the lateral cerebellum when the hand reached for, grasped, raised, then dropped an object (because of the traditional association of the hemisphere with fine movements of the distal extremities), the greatest activity in the lateral cerebellum was indeed with sensory discrimination tasks. These results are relevant to the elaborate patterns of complex food processing found in gorillas, for example. Foods commonly eaten by the mountain gorilla are "protected" by such things as stings, prickles, hard casings, or tiny hooks. Gorilla feeding strategy relies on dextrous movements in a series of logical, hierarchically ordered steps in order to process the food for eating (Byrne, 1995). By implication, the interplay between the sensory and motor in the fine discriminating movements required to process the food would be an activity with considerable cerebellar involvement. The sensory feedback of the lateral cerebellum would of course be important in successful extractive foraging characteristic of all apes and humans.

Great apes have an ability to learn human-based language systems such as Ameslan or Yerkish and an ability to make and use tools under laboratory conditions that far outstrips the performance of monkeys. The plastic-chip system of Premack (1972), the computer lexigrams of Rumbaugh (Rumbaugh, *et al.*, 1973), and American Sign Language (Gardner and Gardner, 1969; Miles, 1983; Patterson, 1978; Patterson and Gordon, this volume; Shapiro and Galdikas, 1999; Terrace, *et al.*, 1979) are all visually based. They require the focus of visual attention and the recognition of important contextual changes, attributes of the lateral cerebellum shown by Allen and colleagues (1997). The inability to shift attention rapidly, particularly between visual and auditory modalities, is an impairment common to both cerebellar and autistic patients (Courchesne, *et al.*, 1994), in keeping with Murakami, *et al.*'s (1989) finding of reduced cerebellar hemisphere size in autistic patients. The focus of attention on relevant cues in an environment would be essential to the learning process, and the prolongation of immaturity in apes relative to monkeys would suggest a greater capacity for learning at crucial maturational stages. The lateral cerebellum, with its contribution to attention and focus and its involvement in procedural learning, could be a much more important element in ape than in monkey cognition.

Gibbons cluster quite clearly with the rest of the hominoids in the lateral cerebellar grade shift. They are often neglected in discussions of ape cognition, being high strung and difficult to test in captivity (Rumbaugh and McCormack, 1966). Given the right conditions, could they learn the same proto-human activities performed by the great apes? The complexity of movement and feeding patterns in the small apes could be tested under controlled conditions as manifestations of visuo-spatial problem solving and other cerebellar-influenced cognitive processes. This is a question to be explored, with the understanding that the rest of the hominoids have experienced another great leap in the evolution of the brain, an increase in absolute size. The great apes and humans have shown both a change in proportions and by implication organization, and a significant increase in quantity of brain tissue. It is the combination of the two steps in evolution that has lead to the outstanding cognitive abilities of the great apes and humans.

The cerebellum has a deep interaction with the neocortex, and so it should not be surprising that it participates in a number of cognitive activities, even more than those outlined above. Our study provides evidence that there was a significant increase in the lateral cerebellum and an important change in the equilibrium of cerebellar circuitry in the hominoid line. The hemisphere to vermis ratio shows a clear grouping of hominoid to monkey and implies a qualitative change in cognitive abilities for gibbons, great apes, and humans. How those structural changes may have affected hominoid cognition can only be suggested here, but future work in the laboratory probing specific cerebellar functions could increase our understanding of the play of the cerebellum in the hominoid mind.

ACKNOWLEDGMENTS

Access to the Yerkes sample was generously provided by Drs. James Rilling and Tom Insel. This project was funded in part by the Leakey Foundation, Simon Fraser University, and the Langara Research Committee, with invaluable support from Murray Besler (Langara), and the Institut für Hirnforschung in Duesseldorf, Germany.

REFERENCES

Akshoomoff, N.A. and Courchesne, E., 1992, A new role for the cerebellum in cognitive operations, *Behavioral Neuroscience* 106(5): 731-738.
Albus, J.S., 1971, A theory of cerebellar function, *Mathematical Biosciences* 10: 25-61.

Allen, G., Buxton, R.B., Wong, E.C., Courchesne E., 1997, Attentional activation of the cerebellum independent of motor involvement, *Science* 275: 1940-1943.

Benefit, B.R., 1999, *Victoriapithecus*: The key to Old World monkey and Catarrhine origins, *Evol. Anth.* 7(5): 155-174.

Bower, J.M., 1997, Is the cerebellum sensory for motors sake, or motor for sensorys sake: The view from the whiskers of a rat. Pp. 463-498 in: (Eds. C.E. de Zeeuw, P. Strata, and J. Voogd), *The Cerebellum: From Structure to Control*, Amsterdam: Elsevier.

Byrne, R., 1995, *The Thinking Ape: Evolutionary Origins of Intelligence*, Oxford: Oxford University Press.

Clutton-Brock, T.H. and Harvey, P.H., 1979, Home range size, population density and phylogeny in primates. Pp. 201-214 in: (Eds. I.S. Berstein and E.O. Smith), *Primate Ecology and Human Origins*, New York: Garland Press.

Courchesne, E., Townsend, J., Akshoomoff, N.A., Saitoh, O., Yeung-Courchesne, R., Lincoln, A.J., James, H.E., Haas, R.H., Schreibman, L., and Lau, L., 1994, Impairment in shifting attention in autistic and cerebellar patients, *Behavioral Neuroscience* 108(5): 848-865.

Doyon, J., 1997, Skill learning, *Int. Rev. Neurobiology* 41: 273-294.

Fiez, J.A., Petersen, S.E., Cheney, M.K., and Raichle, M.E., 1992, Impaired non-motor learning and error detection associated with cerebellar damage, *Brain* 115: 155-178.

Fiez, J.A. and Raichle, M.E., 1997, Linguistic processing. Pp. 233-254 in: (Ed. J.D. Schmahmann), *The Cerebellum and Cognition (International Review of Neurobiology*, vol. 41), San Diego: Academic Press.

Finlay, B.L. and Darlington, R.B., 1995, Linked regularities in the development and evolution of mammalian brains, *Science* 268: 1578-1584.

Fleagle, J.C., 1999, *Primate Adaptation and Evolution, second ed.*, Toronto: Academic Press.

Gao, J.H., Parsons, L.M., Bower, J.M., Xiong, J., Li, J., Fox, P.T., 1996, Cerebellum implicated in sensory acquisition and discrimination rather than motor control, *Science* 272: 545-547.

Gardner, R.A. and Gardner, B.T., 1969, Teaching sign language to a chimpanzee, *Science* 165: 664-672.

Gibson, K.R., 1986, Cognition, brain size and the extraction of embedded food resources, Pp. 93-104 in: (Eds. J.G. Else and P.C. Lee), *Primate Ontogeny, Cognition and Social Behaviour*, Vol. 3, Cambridge: Cambridge University Press.

Kim, S.G., Ugurbil, K., and Strick, P.L., 1994, Activation of a cerebellar output nucleus during cognitive processing, *Science* 265: 949-951.

Leiner, H.C., Leiner, A.L., and Dow, R.S., 1986, Does the cerebellum contribute to mental skills?, *Behavioral Neuroscience* 100(4): 443-454.

Leiner, H.C., Leiner, A.L., and Dow, R.S., 1989, Reappraising the cerebellum: What does the hindbrain contribute to the forebrain, *Behavioral Neuroscience* 103(5): 998-1008.

Leiner, H.C., Leiner, A.L., and Dow, R.S., 1991, The human cerebro-cerebellar system: Its computing, cognitive, and language skills, *Behavioural Brain Research* 44: 113-128.

Leiner, H.C., Leiner, A.L., and Dow, R.S., 1993, Cognitive and language functions of the human cerebellum, *TINS* 16(11): 444-447.

MacLeod, C.E., 2000, *The Cerebellum and Its Part in the Evolution of the Hominoid Brain*. Ph.D. dissertation, Simon Fraser University, Burnaby, BC, Canada.

Marr, D., 1969, A theory of cerebellar cortex, *J. of Physiol.* 202: 437-470.

Martin, R.D., 1980, Adaptation and body size in primates, *Z. Morphol. Anthropol.* 71: 115-124.

Matano, S., 1992, A comparative neuroprimatological study on the inferior olivary nuclei (from the Stephan's Collection), *J. Anthrop. Soc. Nippon* 100(1): 69-82.

Matano, S., Stephan, H., and Baron, G., 1985a, Volume comparisons in the cerebellar complex of primates I. Ventral Pons, *Folia Primatologica* 44: 171-181.

Matano, S., Stephan, H., and Baron, G., 1985b, Volume comparisons in the cerebellar complex of primates II. Cerebellar nuclei, *Folia Primatologica* 44: 182-203.

Matano, S. and Hirasaki, E., 1997, Volumetric comparisons in the cerebellar complex of anthropoids, with special reference to locomotor types, *AJPA* 103: 173-183.

Middleton, F.A. and Strick, P.L., 1994, Anatomical evidence for cerebellar and basal ganglia involvement in higher cognitive function, *Science* 266: 458-461.

Middleton, F.A. and Strick, P.L., 1997a, Dentate output channels: Motor and cognitive components. Pp. 553-568 in: (Eds. C.I. DeZeeuw, P. Strata, and J. Voogd), *The Cerebellum: From Structure to Control* (Progress in Brain Research, Vol. 114), Amsterdam: Elsevier.

Middleton, F.A. and Strick, P.L., 1997b, Cerebellar output channels. Pp. 61-82 in: (Ed. J.D. Schmahmann), *The Cerebellum and Cognition* (*International Review of Neurobiology*, vol. 41), San Diego: Academic Press.

Miles, H.L., 1983, Apes and language: The search for communicative competence. Pp. 43-61 in: (Eds. J. de Luce and H.T. Wilder), *Language in Primates*, New York: Springer-Verlag.

Milton, K., 1981, Distribution patterns of tropical plant foods as an evolutionary stimulus to primate mental development, *Amer. Anthropol.* 83: 534-548.

Molinari, M., Petrosini, I., and Grammaldo, L.G., 1997, Spatial event processing, *Inter. Rev. of Neurobiology* 41: 217-230.

Murakami, J.W., Courchesne, E., Press, G.A., Yeung-Courchesne, R., and Hesselink, J.R., 1989, Reduced cerebellar hemisphere size and its relationship to vermal hypoplasia in autism, *Arch. Neurol.* 46: 689-694.

Parker, S.T. and Gibson, K.R., 1977, Object manipulation, tool use and sensorimotor intelligence as feeding adaptations in Cebus monkeys and great apes, *J. of Hum. Evol* 6: 623-641.

Parker, S.T. and Gibson, K.R., 1979, A developmental model for the evolution of language and intelligence in early hominids, *The Behavioral and Brain Sciences* 2: 367-408.

Patterson, F.G., 1978, The gestures of a gorilla: Language acquisition in another pongid, *Brain and Language* 5: 72-97.

Patterson, F.G.P., and Gordon, W., this volume

Povinelli, D. and Cant, J.G.H., 1995, Arboreal clambering and the evolution of self-conception, *The Quart. Rev. of Biol.* 70(4): 393-421.

Premack, D., 1972, Language in chimpanzees, *Science* 172: 808-822.

Raichle, M.E., Fiez, J.A., Videen, T.O., MacLeod, A-M.K., Pardo, J.V., Fox, P.T., and Petersen, S.E., 1994, Practice-related changes in human brain functional anatomy during nonmotor learning, *Cerebral Cortex* 4: 8-26.

Rilling, J.K. and Insel, T.R., 1998, Evolution of the cerebellum in primates: Differences in relative volume among monkeys, apes and humans, *Brain, Behavior and Evolution* 52: 308-314.

Rumbaugh, D.M. and McCormack, C., 1967, The learning skills of primate: A comparative study of apes and monkeys. Pp. 289-306, in: (Eds. D. Starck, R. Schneider, and H.J. Kuhn), *Neue Ergebnisse der Primatologie (Progress in Primatology)*, Stuttgart: Gustav Fischer Verlag.

Rumbaugh, D., Gill, T., and von Glasersfeld, E., 1973, Reading and sentence completion by a chimpanzee (*Pan*), *Science* 182: 731-733.

Schmahmann, J.D., 1991, An emerging concept: The cerebellar contribution to higher function, *Arch. Neurol.* 48: 1178-1187.

Schmahmann, J.D., 1996, From movement to thought: Anatomic substrates of the cerebellar contribution to cognitive processing, *Human Brain Mapping* 4: 174-198.

Schmahmann, J.D. and Pandya, D.N., 1995, Prefrontal cortex projections to the basilar pons in rhesus monkey: Implications for the cerebellar contribution to higher function, *Neuroscience Letters* 199: 175-178.

Shapiro, G.L., Galdikas, B.M.F., 1999, Sign learning by an adult free-ranging orangutan. Pp. 212-237 in: (Eds. S. Parker, G. Mitchell, and L. Miles), *The Mentalities of Gorillas and Orangutans,* Cambridge: Cambridge University Press.

Stein, J.F., Miall, R.C., and Weir, D.J., 1987, The role of the cerebellum in the visual guidance of movement. Pp. 175-192 in: (Eds. M. Glickstein, C. Yeo, and J. Stein), *Cerebellum and Neuronal Plasticity,* New York: Plenum Press.

Talairach, J. and Tournoux, P., 1988, *Co-planar Stereotaxic Atlas of the Human Brain.* Stuttgart: Thieme Verlag.

Terrace, H.S., Petitto, L., Sanders, R.J., and Bever, T., 1979, Can an ape create a sentence?, *Science* 206: 891-902.

Thach, W.T., 1996, On the specific role of the cerebellum in motor learning and cognition: Clues from PET activation and lesion studies in man, *Behavioral and Brain Sciences* 19: 411-431.

Thach, W.T., 1997, Contex-response linkage. Pp. 599-611 in: (Ed. J.D. Schmahmann), *The Cerebellum and Cognition (International Review of Neurobiology,* Vol. 41), San Diego: Academic Press.

Voogd, J., Feirabend, H.K.P., and Schoen, J.H.R., 1990, Cerebellum and precerebellar nuclei. Pp. 321-386 in: (Ed. G. Paxinos), *The Human Nervous System,* San Diego: Academic Press, Inc.

Welker, W., 1987, Spatial organization of somatosensory projections to granule cell cerebellar cortex: Functional and connectional implications of fractured somatotopy (summary of Wisconsin studies). Pp. 239-280 in: (Ed. J.S. King), *New Concepts in Cerebellar Neurobiology,* New York: Alan R. Liss, Inc.

Zilles, K., Schleicher, A., and Pehlemann, F-W, 1982, How many sections must be measured in order to reconstruct the volume of a structure using serial sections?, *Microscopica Acta* 86(4): 339-346.

SECTION TWO
BONOBOS, THE "FORGOTTEN APE"?
INTRODUCTION

The bonobo (*Pan paniscus*) is arguably the least known African ape (Kano, 1992). Endemic to the Congo River Basin, the bonobo was recognized as a distinct species in the 1930s and was not studied in the wild until the 1970s (de Waal, 2001; Zihlman, 1997). Censuses of wild populations, although crucial to the development of conservation plans, have proved difficult to carry out, in part because of the civil unrest that plagues the region. In 1973, Kano estimated that 100,000 bonobos remained in the wild; by 1992, primatologists hypothesized that this number had dropped to 10,000 to 20,000 individuals. As recently as the mid-1990s, it appeared that not only was the bonobo the "forgotten ape" (de Waal and Lanting, 1997), it was also very likely to go extinct. The three chapters in Section Two help to rectify the dearth of information on the bonobo's status in its natural habitat and give some cause to hope that the species will survive *if* immediate conservation efforts are made on its behalf.

In chapter four, Jef Dupain and L. Van Elsacker review the censuses of bonobos conducted to date and note that the inaccurate replication of Kano's original figures have painted an extremely pessimistic picture. Instead, there may be as many as 50,000 bonobos left in the wild. This reflects a loss of one half the bonobo population estimated by Kano, but is not nearly as low as the 10,000 to 20,000 individuals estimated by the Bonobo/Pygmy Chimpanzee Fund (Japan). Dupain and Van Elsacker describe several major threats to bonobo survival, including commercial and subsistence hunting. On a more optimistic note, they summarize the works of the many people dedicated to this ape's survival. Against a backdrop of an ever-changing government and continued instability in the region, the long-standing commitment to this ape's survival by field researchers, conservationists, and some government officials continues.

JoAnne Myers Thompson describes the Lukuru Wildlife Research Project, which potentially adds to primatologists' understanding of bonobo ecological adaptations. Thompson and her colleagues have found that bonobos in the Salonga National Park occupy a more seasonal environment than is true of other bonobo populations; thus, early bonobos may have foraged "on grassland fruits in a mosaic ancestral habitat". Thompson discusses the formation of a sanctuary at Salonga National Park and the cooperative nature of her work there with the indigenous people, government officials, and outside interests such as logging companies. Of theoretical interest is her observation that several bonobo populations at Salonga National Park have lived in geographic isolation for some time; perhaps we will eventually recognize subspecies within this species. Like the research described by Dupain and Van Elsacker, Thompson's efforts show that despite the impoverished status of the indigenous people, it is possible to protect bonobos and to work with government officials on the species' behalf.

Chie Hashimoto and Takeshi Furuichi present a census of the bonobos inhabiting the Luo Scientific Reserve. Species such as chimpanzees (*Pan troglodytes*) and orangutans (*Pongo pygmaeus*) have been censused using nest counts, and Hashimoto and Furuichi test the efficacy of this technique for bonobos. They census bonobo populations living in northern and southern regions of the reserve. To the north, they compare population density results obtained via nest counts with estimates derived from known groups and their ranging habits. These data are then used to estimate bonobo population density in the south, where no groups have yet been habituated. Bonobo density in the southern part of the reserve is approximately half the density measured to the north. They note that this cannot be explained by differences in human predation pressures on bonobos. Instead, they suggest that differences in vegetation might result in higher densities to the north. The results described by Hashimoto, Furuichi, and Thompson indicate that considerable research is needed to better understand bonobo ecology and distribution.

The three chapters in Section Two demonstrate that bonobos, while in a precarious situation, remain in numbers sufficient for the species' longterm survival. As Dupain and Van Elsacker note, "while the species' situation is urgent, it is not yet hopeless".

REFERENCES

de Waal, F.B.M. and Lanting, F., 1997, *Bonobo: The Forgotten Ape*, Berkeley: University of California Press.

Kano, T., 1992, *The Last Ape: Pygmy Chimpanzee Behavior and Ecology*. Stanford, CA: Stanford University Press.

de Waal, F.B.M., 2001, Apes from Venus: Bonobos and human social evolution. Pp. 41-68, in: (Ed., F.B.M. de Waal), *Tree of Origin: What Primate Behavior Can Tell Us About Human Social Evolution*, Cambridge, MA: Harvard University Press.

Zihlman, A., 1997, African apes. Pp. 17-23, in: (Ed. F. Spencer), *History of Physical Anthropology, Volume 1 (A-L)*, New York: Garland Publishing, Inc.

Chapter 4

THE STATUS OF THE BONOBO (*PAN PANISCUS*) IN THE DEMOCRATIC REPUBLIC OF CONGO

J. Dupain and L. Van Elsacker
Royal Zoological Society of Antwerp, Kon. Astridplein 26, B-2018 Antwerp, Belgium

1.   INTRODUCTION

In this chapter we provide an overview of the bonobo's distribution, the status of the species over time, threats to its survival, and current activities of researchers and conservationists. Such a review is easier for the bonobo than it would be for the chimpanzee (*Pan troglodytes*). The bonobo consists of one species (*P. paniscus*) with no recognized subspecies. It is a species that inhabits only one country, and whose distribution is restricted to a smaller variety of habitats in contrast to the countries and varied habitats of the chimpanzee.

The most detailed existing report on the status of free-ranging bonobos is the last *Action Plan for Pan paniscus: Report on Free Ranging Populations and Proposals for their Preservation* (Thompson-Handler, *et al.*, 1995). The plan is a compilation of all information available through 1993. It reports on the situation at existing and former research sites and lists priority actions needed in the strategy for the conservation of the wild bonobo population in the former Zaire. Unfortunately, very few of the proposed actions were since undertaken. Subsequently, the situation in the field changed dramatically. Zaire became the Democratic Republic of Congo (DRC) in 1997. The beginning of 1998 was characterized by new hope for conservation, and field researchers resumed activities.

Despite its restricted distribution, surprisingly little is known about the bonobo. The first major publications on wild bonobos date from the 1980s (Badrian and Badrian, 1977; Badrian, *et al.*, 1981; Horn, 1980, Kano, 1982,

1983, 1984, 1986; Susman, 1984), and since that time the need for more research, especially surveys, is emphasized at nearly every meeting (Lee, *et al.*, 1988; Malenky, *et al.*, 1989; Thompson-Handler, *et al.*, 1995; Van Elsacker, *et al.*, 1996; Van Elsacker, *et al.*, 1997). Simultaneously, researchers urge the implementation of conservation actions. Unfortunately, political instability, insecurity, uncertainty, and war were used to postpone new investments and curtail the implementation of existing conservation plans. These problems did not halt the decimation of free-ranging bonobo populations. Hunting pressure is growing, and habitat destruction continues. One could wonder whether this chapter is yet another attempt to characterize the bonobo's precarious situation and re-emphasize the need for conservation action on the species' behalf. Although such a description is in part unavoidable, we hope that it will become clear that there is little justification to postpone necessary action for the sake of the bonobo, and that a small group of people has been and still is active in various respects. These people know that if they do not labor on behalf of the bonobos no one will.

2.   BONOBO DISTRIBUTION AND NUMBERS

The distribution of the bonobo is restricted to the south bank of the Congo River (Figure 12). To the south, its geographical range is confined by the Kasaï-Sankuru Rivers. Some researchers consider the most likely eastern boundary to be the Lomami River (Coolidge, 1933; Kano, 1992); others note that bonobos are present in the "Lomami-Lualaba forest" (Colyn, 1987; Kortlandt, 1995; van den Audenaerde, 1984). Whether or not the latter forest block is included, the area of bonobo's potential distribution covers more than 500,000 km² (Kano, 1984). Based on historical information, Kortlandt (1976) estimated the actual range within this potential distribution area to be 350,000 km², but he did not take into account the patchiness of the species' distribution (Kano, 1984). Using the results of the only extensive survey conducted to date, Kano (1992) estimated that the bonobo population was probably limited to a geographical range of 135,000 km², situated in the northern part of the species' potential range (Figure 12). He feared the extinction of the bonobo in the southern part, and thus felt the total range could not exceed 200,000 km². More recent surveys have confirmed the presence of bonobos in the southern range (Thompson-Handler, *et al.*, 1995). Unfortunately, the first publication of Kano's survey (1984) contained a typographical error and referred to the estimated bonobo range as being 13,500 km² (not 135,000 km²). Surprisingly, Thompson-Handler, *et al.* (1995) did not question this error. In all recent publications that mention wild bonobo distribution range, one or other of these estimates is noted, and the uncertainty of the figures is rarely stressed.

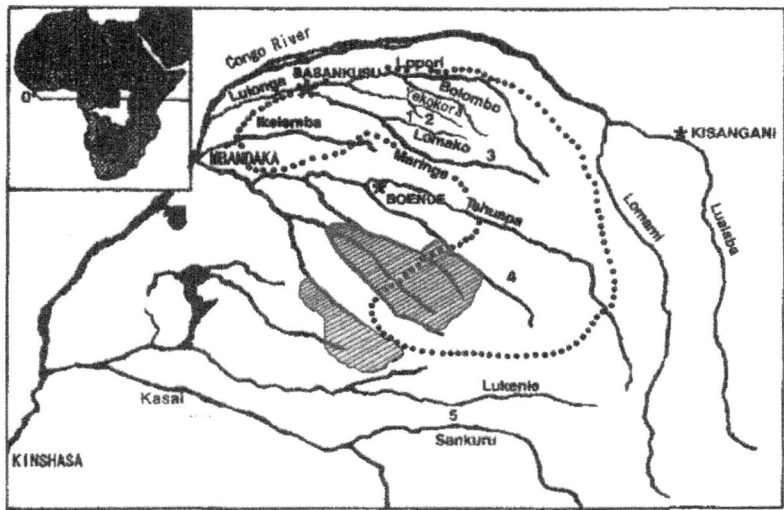

Figure 12. The distribution of the bonobo is restricted to the south bank of the Congo River (previously known as the Zaire River).
Legend:
Heavy black line: Estimated range of bonobos to the north (Kano, 1984).
Shaded area: Salonga National Park
Research sites: 1: Iyema-Lomako (Royal Zoological Society of Antwerp); 2: LFPCP (Stony Brook University); Isamondje-Lomako (Max Planck Institute); 3: Wamba (Kyoto University); 4: Yalosidi (Kyoto University); 5: Lukuru Wildlife Research Project (University of Oxford).

With the lack of concrete evidence as to the exact distribution pattern of the bonobo, one can imagine that little research has been conducted to explain the patchiness of the species' distribution. Human impact on habitat, vegetation patterns, and seasonal flooding of the forest likely influence distribution (Kano, 1984, 1992; Kortlandt, 1995), but more surveys on both bonobo densities and vegetation patterns are needed to understand this.

When it comes to estimating the total bonobo population, the need to conduct more surveys and compile available information becomes even more striking. All available estimates are based on the single and most extensive survey conducted by Kano in 1973. As Kano (1984) pointed out, he arrived arbitrarily at an average population density of 0.4 ind/km², hence his estimate of 54,000 bonobos total in the northern range (135,000 km²). This suggested a total population of less than 100,000 individuals; following Kano's rationale, the existence of a bonobo population in the southern range would have brought the total number to 100,000 or more. Kortlandt (1995), relying on Kano's estimate, also reported a total of 100,000 bonobos. The *Action Plan for Pan paniscus* (Thompson-Handler, *et al.*, 1995) quoted Kano's incorrectly printed original figure for the northern distribution range, and thereby estimated an overall population size of 5,000 individuals or less! This figure was based on four errors (Kortlandt, 1996) but is generally used to emphasize the endangered status of the bonobo.

In 1992, the Bonobo/Pygmy Chimpanzee Protection Fund (Japan) reported that the bonobo population had been reduced to half its original size over a 16-year period. This estimate was based on the 50% reduction of bonobo habitat over 16 years at the Wamba research site, and the virtual disappearance of once large numbers of bonobos at locations such as Yalosidi. A new species total of 10,000 to 20,000 individuals was proposed. Today, most scientists refer to this range when speaking of the bonobo's status in the wild. One should be aware of the fact that all existing estimates are based on Kano's publications (1984, 1992), in which the author emphasizes the tentativeness of his figures, and on the estimated 50% reduction of the bonobo population over 16 years (Bonobo/Pygmy Chimpanzee Protection Fund (Japan), 1992).

More recent information suggests that this figure of 20,000 individuals is too low. In 1992, a total of about 3,000 animals were estimated to live in the 6,000 km² Upper Luo Region (encompassing the Wamba research site), which was a proposed protected area (Bonobo/Pygmy Chimpanzee Protection Fund (Japan), 1992) (Figure 12). We estimate through our own fieldwork that the proposed Lomako Reserve (Figure 12), with an area of about 3,800 km², still contains about 2,000-3,000 individuals. Based on the first disappointing surveys (e.g., Kano, 1979), doubts about the presence of bonobos at Salonga National Park (36,000 km²) were expressed. However, later evidence indicates that bonobos are present at Salonga, but their status is unknown (historical records: see Kortlandt, 1995; van den Audenaerde, 1984; surveys: Meder, *et al.*, 1988; Reinartz, pers. comm. to J.D.; pers. comm.: Colyn, missionaries, owners of plantations). While surveying elephants at Salonga, Alers, *et al.* (1989) occasionally encountered bonobos and reported that they probably exist throughout the park but with a patchy distribution (see East, 1990). If one assumes, for heuristic purposes, a density of 0.4 ind/km² as proposed by Kano (1984) over the entire park, Salonga may harbor a total population of 14,400. Essential bonobo populations live south of the Salonga National Park. Boats traveling down the Lukenie River invariably transport bonobo orphans to Kinshasa (Dupain, unpublished information). Additionally, bonobos are present between the Lukenie and

Sankuru Rivers (Thompson, this volume). A survey carried out in the area southwest of the Salonga National Park confirmed the presence of bonobo populations (Thompson, this volume; Thompson-Handler, *et al.*, 1995). Kano (1984) recorded the presence of bonobos in the southern part of the forest block between the Lopori and Bolombo Rivers. The middle and northern parts of this forest block also seem to have larger bonobo populations (Dupain, unpublished information). According to some commercial hunters (Dupain, unpublished information), Kano (1984), and Kuroda (pers. comm. in Thompson-Handler, *et al.*, 1995) the same holds true for the forest at the headwaters of the Lopori River. Kortlandt (1995) confirms the presence of bonobos in the area of the upper Tshuapa and the tributaries of the Lomami. The Mbandaka-Boende-Basankusu triangle, intersected by the Ikelemba River, is most probably also home to bonobo populations. Kano (1984) confirms their presence here both by direct observation and indirect information. Bonobos are apparently distributed both south and north of the Ikelemba River (Dupain, unpublished information; Kortlandt, 1995; Thompson-Handler, *et al.*, 1995). Finally, except for some historical records (Kano, 1984; Kortlandt, 1995), no data are available for the forest block between the Lulonga-Lopori Rivers and the Congo River. Putting all this information together, one can deduce tentatively that an overall bonobo population of about 50,000 individuals or more may still be living in the Central Basin. This is half the estimate made by Kano (1994) and Kortlandt (1995).

This is indeed a positive note compared to the widely accepted 10,000-20,000 figure. However, we should be cautious in extrapolating from such limited information. In 1990, World Wildlife Fund-International proposed the gazetting of the 3,800 km² Lomako Reserve (Figure 13). Based on their information, it was believed that there were no permanent indigenous communities living along the Lomako River or north of the established bonobo research sites (based at the southern edge of the proposed reserve) (Thompson-Handler, *et al.*, 1995). In 1994 and 1995 we estimated, based on nest counts, a density of about 2 bonobos/km² living in a research area along the Lomako River (Dupain and Van Elsacker, this volume; Dupain and Van Krunkelsven, 1994; Dupain, *et al.*, 1996). This figure corresponds with densities at adjacent research sites. Supposing an even distribution throughout this forest block (but see Kano, 1984), one may tentatively assume that 6,000 bonobos live within the proposed Lomako Reserve. However, very high hunting pressure was witnessed during a one and a half month survey (conducted in 1995) in this "undisturbed" forest block along the Yekokora River (Dupain and Van Elsacker, this volume; Dupain, *et al.*, 2000) and most of the formerly dense bonobo population in the northern part of the block appears to have been decimated. Therefore, a more accurate estimate may be the 2,000-3,000 individuals mentioned previously. Not one field researcher, although working close to this area, was aware of the severe hunting pressure on this bonobo population. Likewise, we are probably unaware of what is happening in most of the Salonga National Park, between the Lomami and Lualaba Rivers, between the Lulonga and the Ikelemba Rivers, along the Bolombo River, and so on.

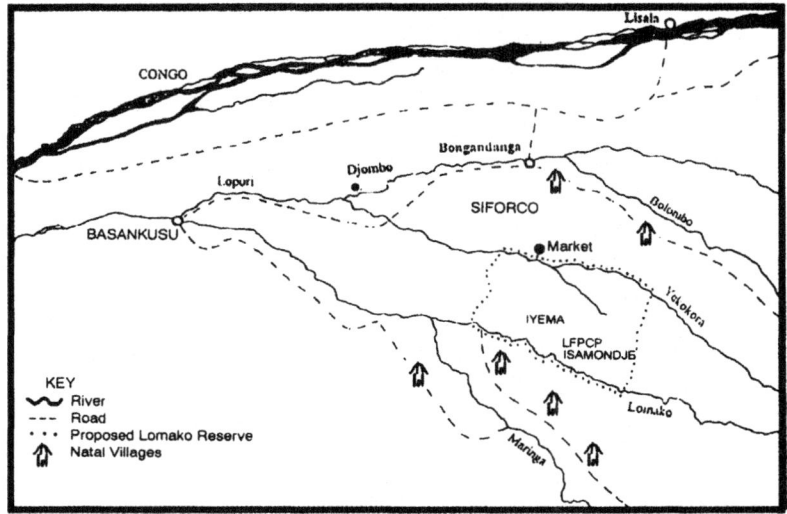

Figure 13. The Lomako Forest

Potential threats to the total bonobo population are similar to the alarming reports of the slaughter of chimpanzees and gorillas elsewhere in Central Africa. Once these areas were protected by their inaccessibility, but this is no longer true. The same problems confront conservationists of other African primates. An increase in commercial hunting is a consequence of deteriorating governmental infrastructure, facilitated by the opening up of unexploited forest along logging roads and the lack of law enforcement. We highlight here some of these major threats confronting bonobo populations.

3. THE MAJOR THREATS TO BONOBO SURVIVAL

3.1 Logging

Most of the forests that still have bonobo populations are subdivided into logging concessions. Nearly the entire forest block between the Yekokora and Bolombo Rivers belongs to the logging company Siforco, the forest block between the Bolombo and Lopori Rivers, north of Djolu, is divided into two concessions, and other companies have concessions on almost all of the forest between the Ikelemba and the Lulonga Rivers, the forest between the Lulonga-Lopori and the Congo Rivers, etc. (Dupain, unpublished information). It is estimated that over 55% of the current bonobo range is either covered by active concessions or standing permits (Reinartz and Inogwabini, pers. comm.). The fact that 86% of the original forest cover remains is due to the reluctance of foreign companies to invest in the DRC and the deteriorating infastructure of the DRC government (Bryant, *et al.*, 1997). However, the present government sees the forests as an important source of revenue and one that can be realized fairly quickly within the next few years (Inogwabini and Reinartz, pers. comm.). Selective logging is practiced. Yet the indirect impact of the timber trade is dramatic, and because it facilitates access into the forest and human and bushmeat transport, it creates a market for bushmeat. Some lumber company owners argue that only traditional, subsistence hunting occurs in their concession, and that there is no transport of bushmeat, infant primates, and rifle cartridges on their boats. Others admit that they pave the way for the bushmeat and pet trades and that consequently the forests are depleted of wildlife. We use here the sequence of events at Lomako Forest to illustrate the typical connection between logging and the bushmeat trade.

Originally, most of the forest block between the Maringa-Lopori-Bolomba-Lomako Rivers belonged to the lumber company Siforco (Danzer, Furnierwerke, GMBH & Co). Siforco is the biggest logging company active in the DRC, with concessions covering a total of 2.7 million ha (Wolfire, *et al.*, 1998). In 1987, Siforco halted their activities in the forest block between the Lomako and Yekokora Rivers and ceded their wharf on the Maringa River to the management of World Wildlife Fund-Germany. Currently, Siforco is still active north of the Yekokora River, an area that is north of our field site, Iyema-Lomako (Figure 13). This entire forest block, south of Bongandanga, traditionally belonged to the Mongo people (Van der Kerken, 1944). These people were agriculturalists. Hunting in the Lomako Forest was mostly for subsistence, and eating bonobos was taboo, so the Lomako Forest was not heavily hunted.

About 200 employees and their families, many of them immigrants, live at Siforco's site north of the Yekokora River. No domestic meat is provided to them, so they hunt or buy bushmeat. An increased demand for animal protein has created a market that attracts other hunters, who have migrated into the local Mongos' forest. A hidden market that is the meeting place of lumber company employees and local and immigrant hunters was discovered during our 1995 survey (Dupain, *et al.*, 2000). Here bushmeat and fish are exchanged for clothes, medicines, soap, sugar, rifle cartridges, steel cables, and other necessities. Siforco employees and their family members are consumers of, and increasingly important traders in, the bushmeat economy. They use the lumber company boats to import textiles, medicines, and rifle cartridges and to export smoked bushmeat to major cities where the demand for protein is high. This trade is a major cause of the depletion of the fauna (including bonobos) in and around logging concessions. When the company leaves the concession, a rootless group of unemployed people is left behind. Commercial hunting is one of the solutions to their desperate plight. The risk of an almost complete defaunation of the area is very high.

The effects of selective logging are not obvious. From the air, the influence is negligible. When compared to other countries such as Gabon and Cameroon, the export of tropical wood from the DRC is minimal (in 1995 DRC: 300,000 m^3, Cameroon: 2.7 $10^6 m^3$, Gabon: 2.1 $10^6 m^3$; Organisation Internationale des Bois Tropicaux, 1997 in Wolfire, *et al.*, 1998). Very few people are aware of what is happening in the concession areas and the surrounding regions. However, if one spends some time in these remote places, one discovers that numerous bonobo populations are being decimated at an increasing pace by enterprising people surviving under most difficult circumstances.

3.2   Forced to Hunt

Although the presence of logging companies creates and organizes market opportunities and attracts external hunters, even in the absence of these companies, the original local population must make a living. The relative importance of hunting in this endeavor will ultimately determine hunting pressure on Africa's fauna. Considering the Lomako area and its traditional owners, the Mongo people (our partners in the Bonobo *in Situ* Project) as a case study, the situation is very straightforward. The Mongo people live in their natal villages along the road south of the Lomako River (Figure 13). Formerly, they could make a living by selling crops from their plantations (mainly coffee, but also cocoa, maize, etc.). About twice a year, small groups of men or families traveled more than 20 km into the forest to hunt for a few weeks. This catch was shared with an entire village. This form of subsistence hunting constituted a minor pressure on bonobo populations.

Today, due to the deteriorating governmental infrastructure of the region, the commercial relationship between the coffee businessmen from Kinshasa and the owners of the plantations (at remote locations in the forest) is broken. The local population cannot predict when or even if they will have the chance to sell their crops. In recent years some families have opted for a less comfortable but more certain way of life: they leave their villages and settle along the river or in the forest to fish or hunt whatever they can (Figure 14). Commercial hunting is becoming a significant contribution to household livelihoods (Hart and Petrides, 1987; Muchaal and Ngandjui, 1995; Wilkie, *et al.*, 1992) (Figure 15). Family members smoke bushmeat and travel to major cities where they exchange it for necessities. The system works, but as a consequence the forest, formerly uninhabitated by humans, becomes subject to ever growing human populations and hunting pressure. This lifestyle will in the near future lead to the depletion of the resources upon which these people depend (see also Dupain, *et al.*, 2000; Noss, 1995; Wilkie, *et al.*, 1998). Conservation strategies will only succeed if alternative solutions are found for the local population to make a living.

Figure 14. People leave their natal villages and settle along the river or in the forest.
© P. Seeuws (Bonobo-in-Situ, RZSA)

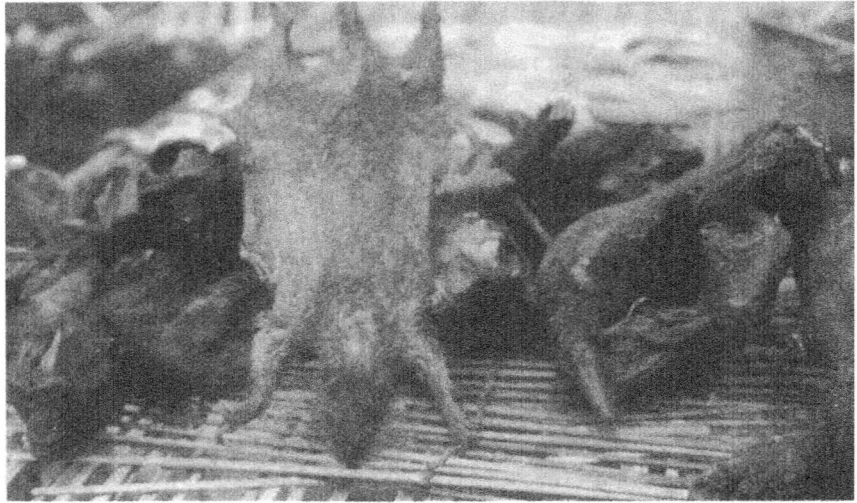

Figure 15. People hunt whatever they can to make a living.
© J. Dupain (Bonobo-in-Situ, RZSA)

3.3   Kinshasa

Following the riots in 1991 and 1993, most expatriates abandoned Zaire. Expatriates were the major market for the pet trade. Although some boat owners confirm that bonobo infants can be found on every boat coming into Kinshasa, the capital no longer constitutes an important pet market. Trade in bonobo infants is more likely a side effect of the bushmeat trade. Nevertheless, conservation education aimed at potential pet-buyers should continue.

With its 5 to 6 million inhabitants, Kinshasa is the largest and most lucrative market for animal protein. One might expect the bushmeat trade to be most obvious there, but this is not the case. Only a minor proportion of the total bushmeat trade is witnessed at the main port in Kinshasa or when visiting the market in the capital's center. Bushmeat traders leave boats with their merchandise at Maluku, about 80 km outside of the capital. They are aware that many of the products they sell are illegal, and they know which animals are officially protected. The intensity and the uncontrolled nature of the bushmeat trade is affirmed by travelers and by boat owners. Although the demand for meat will not disappear, the extent to which this need will be satisfied by the bushmeat trade depends on law enforcement, local economics, and the existence of subsistence alternatives for hunters and traders.

4. CURRENT ACTIVITIES

4.1 Presence of Researchers

The presence of researchers in the field may help to protect local bonobo populations (Dupain, *et al.*, 2000; Thompson-Handler, *et al.*, 1995). If researchers work in collaboration with the local population, the extent of their influence may become even more significant.

Currently, five bonobo research sites exist (Figure 12). The best known of these are the sites at Wamba and Lomako. The research site at Wamba was established in 1974 under the direction of Dr. Kano of Kyoto University (e.g. Hashimoto and Furuichi, this volume; Kano, 1992). The American Lomako Forest Pygmy Chimpanzee Project (LFPCP) was established by Dr. Susman of Stony Brook University in 1980 (e.g. Susman, 1984). Data have not been continuously collected at Lomako. Dr. Hohmann of the Max Planck Institute has since started another bonobo research project at Lomako (e.g. Fruth and Hohmann, 1993). Approximately 20 km from these two sites, we established the Iyema research site in 1995 (Royal Zoological Society of Antwerp, Dupain and Van Elsacker, this volume; Dupain, *et al.*, 1996). In 1992, Dr. Thompson of the University of Oxford established the Lukuru Wildlife Research Project in the Zone Dekese. She collected data on bonobos and their ecology between the Lukenie and Sankuru Rivers (e.g. Thompson, this volume). This site is particularly interesting as the habitat is not continuous dense forest but a mosaic of forest and grassland.

4.2 Bonobo Conservation and Logging

In 1998, Karl Ammann organized an expedition to the north of the Lomako River to document the impact of the nearby Siforco logging concession on the bonobos inhabiting the proposed Lomako Reserve. The evidence collected helped to initiate a dialog with the concession holders (Ammann, 1998; Dupain, unpublished information). As a result, Siforco may assist in the creation of the proposed reserve (see 3.1. above). Meanwhile, there appears to be a general trend for Asian companies to establish new concessions in central Africa, and thus also within the bonobo's range area (Dupain, unpublished information; Thompson, 1998, 1999). Good relations with the DRC government have already proved invaluable in establishing future conservation policies that address the interests of logging companies.

4.3   Socioeconomy and Agriculture

With the support of the Bonobo Protection Fund, an agricultural program is planned for the area around the Lomako River. At the request of the local people, this program aims to assure the sale of local peoples' coffee crops. We hope to reverse the migration process described above; the families who once were forced to migrate into the forest should thereby be able to return to their plantations.   Simultaneously, the relationship between the human populations and the Lomako Forest will be studied.   We will consider the impact of hunters on the forest fauna, the farmers on the plantations near the Lomako Forest, and the inhabitants of Basankusu, who constitute the most important market for products of the Lomako Forest.

4.4   Protected Areas

The above mentioned studies and surveys in the Lomako Forest should culminate in a strong justification for the creation and effective, collaborative protection of the Lomako Reserve.

In 1992, after two decades of research at the Wamba research site, the government of the former Zaire established the Luo Scientific Reserve (481 km²) (Hashimoto and Furuichi, this volume²). At the request of the Bonobo Protection Fund, Dr. Kuroda proposed the creation of the Luo Special Protection Area, which covers 6,000 km² and harbors about 3,000 bonobos (Bonobo/Pygmy Chimpanzee Protection Fund (Japan), 1992). This area has yet to be given official protected status.

The Salonga National Park was created in 1970 and declared a World Heritage Site in 1984.   It is the only federally protected area for *Pan paniscus*. Although the presence of bonobos makes this site of particular interest, knowledge of the area is very poor. The World Conservation Union (IUCN) (1990) reported severe poaching pressure there.   In 1997, the Zoological Society of Milwaukee began a partnership with the Institut Congolais pour la Conservation de la Nature (ICCN) to conduct a three-phase regional survey of the Salonga National Park. The reconnaissance survey was completed in January 1998. After training Congolese researchers during the second phase, an 18-month comprehensive survey of the park will be undertaken (G. Reinartz, pers. comm. to JD).

4.5   Orphaned Bonobos and Education Programs

At the time of our first visit to the former Zaire in 1994, the existence of bonobos was almost unknown to the people living in Kinshasa. Since then much has changed. The joint efforts of Mme. Claudine André's promotion of school and other educational visits at the bonobo sanctuary, Mme. Messenger's educational magazines for children in Lingala, and our lectures on our research in the forest have made the bonobo a well-known figure. After many years of tireless work by Mme André, the president of the not-for-profit association Amis des Animaux du Congo, a temporary sanctuary for confiscated bonobo orphans has been established in the center of Kinshasa, in partnership with the DRC Ministry of Environment (Figure 16). Mme André continues to correspond with people at all levels: Congolese government officials, embassy personnel, businessmen, teachers, and others, thereby exerting a broad spectrum of influence ultimately aimed at the conservation and well-being of the wild bonobo population.

Figure 16. Confiscated bonobo orphans in Kinshasa arrive at the sanctuary created and managed by Amis de Animaux du Congo (Mme. Cl. Andre-Minesi).
© J. Dupain (Bonobo-in-Situ, RZSA)

4.6   The Government

Considerable effort has been expended to implement bonobo conservation plans, and there has been a positive response from the government of the DRC. Currently, the bonobo issue is one of the Ministry of Environment's priorities. All of the above mentioned efforts, research *in situ*, expeditions to logging concessions and dialog with owners, socioeconomic studies of the local population, the creation of the Luo Scientific Reserve, expeditions into the Salonga National Park, the creation of an orphan bonobo sanctuary, and the confiscation of bonobos by government officials in Kinshasa, are supported by the Congolese government. However, these varied activities lack coordination and coherence. To address this issue, and by invitation from the DRC's Ministry of the Environment, a Population Habitat Viability Analysis (PHVA) workshop on bonobos was planned for 2000 (but see Postscript below). All stakeholders were to converge at this workshop to discuss the future of the bonobo.

5.   CONCLUSION

The total bonobo population may be larger than the most conservative estimates referred to in early publications that ignore the likely existence of several as yet unidentified bonobo populations. However, free-ranging bonobo populations are currently being decimated by the Congolese people, who are desperately seeking a way to eke out a living. This devastating impact is caused not by subsistence hunting, but by ever growing commercial hunting. An increasing fragmentation of the remaining bonobo population is a likely outgrowth of this hunting. An aerial view of the forest provides few clues of the slaughter taking place beneath the canopy; only a resident becomes aware of this. The dearth of more detailed figures on bonobo distribution, densities, and ecology and on socioeconomic conditions of local human populations should be emphasized at every opportunity.

Despite a lack of significant investment by donor organizations, a small group of dedicated researchers and conservationists continue their grassroots efforts. We make an appeal to all interested institutions and agencies to resume or increase their support of researchers and conservationists working in the field under sometimes desperate conditions. We must not wait until *Pan paniscus* becomes a critically endangered species and assumes a place next to the mountain gorilla on the IUCN *Red List of Threatened Species*. The bonobo's situation is urgent, but it is not yet too late to intervene on the species' behalf.

6.   POSTSCRIPT (SEPTEMBER 2000)

The war resumed in the DRC only one month following the August, 1998 presentation of an earlier version of this chapter at the Great Apes of the World (GAWC-3) conference. By October 1998, Iyema was the last staffed bonobo research site. We finally left in November 1998. For more than one year now, the front lines of the various factions are situated in the center of the bonobo's range. Despite this conflict, the German research team (led by G. Hohmann of the Max-Planck Institute) succeeded in conducting a first survey in the easternmost part of the bonobo distribution range. None of the existing study sites are currently accessible. We can only guess to what extent the habituated bonobo populations have been affected and whether hunting pressure is growing. During the take-over by President Kabila, people were afraid to travel deep into the forests for personal safety reasons, but this fear was short lived. Now the many people who have been displaced by the war seek refuge, food, and shelter in the forests. Meanwhile, the number of bonobo orphans arriving in Kinshasa is growing: between November 1999 and June 2000, seven bonobos were confiscated by the Amis des Animaux du Congo and the Ministry of Environment, as opposed to an average of one confiscation per year in past years (Mme. André, pers. comm. to JD). Another four infant bonobos were sighted for sale in Kinshasa during the same period. One can expect intensified hunting in habitat areas.

However, more conservation activity than ever before is going on in Kinshasa. Mme André still manages the sanctuary for confiscated bonobos, which is primarily supported through private donations from concerned individuals and organizations. She continues her efforts to discuss the future of the bonobo in the DRC. In June 2000, she organized a workshop with various authorities, including the Congolese armed forces, to increase awareness among relevant field-based cadres and decision-makers of the risk posed to the bonobo by the current conflict situation and the increased trades in bushmeat and pet orphans that result from it. Through collaboration between USAID, the Zoological Society of Milwaukee (ZSM), Bleu Blanc, and the Congo River Baptist Community, 40,000 education leaflets on bonobos will be distributed throughout the conflict zone (Reinartz and Russell, pers. comm. to JD). In the form of emergency aid, the ZSM has recently sent funds to the Institut Congolais pour la Conservation de la Nature to pay park guards and establish anti-poaching river patrols in the Salonga National Park. The PHVA workshop planned for 2000 did not take place, in part due to the lack of information from the field. A bonobo conservation status assessment workshop was organized in November 1999, during the Conference of Excellence (COE) international symposium held in Inuyama (Japan) and facilitated by the Conservation Breeding Specialist Group (CBSG). Many concerned field researchers and conservationists were present. Currently, the possibility of holding a Bonobo Conservation: Urgent Action meeting in June 2001 is being investigated. This would be the first meeting of this kind to be organized in Kinshasa, in the presence of all relevant parties.

Most significantly, important donor bodies have now focused their attention on the DRC. To list these organizations is beyond the scope of this chapter (see USAID, 2000). However, on this positive note in this postscript, we can perhaps hope that peace will return to the DRC, for the sake of the Congolese people and their natural resources.

ACKNOWLEDGMENTS

We thank Gary Shapiro and Biruté Galdikas for their invitation to participate in the Great Apes of the World Conference and to contribute to this volume. We are grateful to the Ministère de l'Environnement, Protection de la Nature et Tourisme and the Ministère de l'Enseignement Supérieur, Recherche Scientifique et Technologie, Kinshasa, République Démocratique du Congo, for the very kindly provided authorizations and mission orders. Fieldwork was made possible by the financial support of the KBC. Further financial support was provided by the Wildlife Conservation Society, the LSB Leakey Foundation, the Fonds Leopold 111 pour l'Exploration et la Conservation de la Nature, and the Bonobo EEP members. All logistical support was provided by the Belgian Embassy and Cooperation in Kinshasa (DRC), CDI-Bwamanda, Claudine André and Pierre Verhaeghe (AAC), Paul DePetter and J.Cl. Hoolans (Nocafex), Father Paul (Procure Saint-Anne, Kinshasa), and the missionaries of Mill Hill (Basankusu). The research would not have been possible without financial and other support from the RZSA. We thank Dir. F.J. Daman for his continuous support of the Bonobo in Situ Project. We are thankful to Claudine André, Sally Cox, Takeshi Furuichi, Chie Hashimoto, Gay Reinartz, Diane Russell, and Jo Thompson for the latest updates on the bonobo's status. We thank Helen Attwater and Lori Sheeran for editorial assistance. The first author was generously supported by the Center of Excellence Research as a visiting research scholar at the Primate Research Institute (Kyoto University) during the writing of this chapter.

REFERENCES

Alers, M., Blom, A., Kiyengo, A., and Masunda, T., 1989, Reconnaissance des elephants des forêts du Zaire. Rapport de Mission (Janvier-Mars 1989). WCI/IZCN/WWF.
Ammann, K., 1998, The conservation status of the bonobo in the one million hectare Siforal/Danzer logging concession in central D.R. Congo. Eletronic document, http://biosynergy.org/bushmeat/, accessed July 1998.
Badrian, A. and Badrian, N., 1977, Pygmy chimpanzees, *Oryx* 13: 463-472.

Badrian, N., Badrian, A., and Susman, R., 1981, Preliminary observations on the feeding behavior of *Pan paniscus* in the Lomako forest of central Zaire, *Primates* 22(2): 173-181.

Bonobo/Pygmy Chimpanzee Protection Fund (Japan), 1992, *A Plan for the Protection of Bonobos (Pygmy Chimpanzees) of the Upper Luo Region.*

Bryant, D., Nielsen, D. and Tangley, L., 1997, *Last Frontier Forests: Ecosystems and Economies on the Edge*, World Resource Institute, Washington DC.

Colyn, M., 1987, Les primates des forêts ombrophiles de la cuvette du Zaire: Interprétations zoogéographiques des modèles de distribution, *Revue de Zoologie Africaine* 101: 183-196.

Coolidge, H., 1933, *Pan paniscus* (pygmy chimpanzee) from south of the Congo River, *Amer. J. Phys. Anthropol.* 8(1): 1-57.

Dupain, J., and Van Elsacker, L., this volume.

Dupain, J. and Van Krunkelsven, E., 1994, A new study site for bonobo (*Pan paniscus*) research. 1st benelux-congress of Zoology, Leuven, Belgium (Poster abstract p. 66).

Dupain, J., Van Krunkelsven, E., Van Elsacker, L., and Verheyen, R.F., 1996, Iyema: A new field site for bonobo (*Pan paniscus*) research. In: Abstract of the 1° Congreso de la Asociación Primatológica Española. APE '96. European Workshop on Primate Research. Madrid, Spain, October 16-18, 1996: 28.

Dupain, J., Van Krunkelsven, E., Van Elsacker, L., and Verheyen, R.F., 2000, Current status of the bonobo (*Pan paniscus*) in the proposed Lomako Reserve (Democratic Republic of Congo), *Biological Conservation* 94: 254-272.

East, R., 1990, *Antelopes, Global Survey and Regional Action Plans. Part 3. West and Central Africa*, IUCN, Gland.

Fruth, B. and Hohmann, G., 1993, Ecological and behavioral aspects of nest building in wild bonobos (*Pan paniscus*), *Ethology* 94: 113-126.

Hart, J.A. and Petrides, G.A., 1987, A study of relationships between Mbuti hunting systems and faunal resources in the Ituri Forest of Zaire. Pp. 12-15 in: *People and the Tropical Forest*, US Department of States, US Man and Biosphere Program, Washington, DC.

Hashimoto, C. and Furuichi, T., this volume.

Horn, A., 1980, Some observations on the ecology of the bonobo chimpanzee (*Pan paniscus*, Schwarz 1929) near Lake Tumba, Zaire, *Folia Primatologica* 34: 145-169.

IUCN, 1990, *La conservation des écostystèmes forestiers du Zaire*. IUCN, Gland, Switzerland and Cambridge.

Kano, T., 1979, A pilot study on the ecology of pygmy chimpanzees (*Pan paniscus*) of Yaalosidi, Republic of Zaire, *Int. J. Primatology* 4: 1-31.

Kano, T., 1982, Social group of pygmy chimpanzees (*Pan paniscus*) of Wamba, *Primates* 23: 171-188.

Kano, T., 1983, An ecological study of the pygmy chimpanzees (*Pan paniscus*) of Yalosidi, Republic of Zaire, *Int. J. Primatology* 4: 1-31.

Kano, T., 1984, Distribution of pygmy chimpanzees (*Pan paniscus*) in the central Zaire Basin, *Folia Primatologica* 43: 36-52.

Kano,T., 1986, *Saigo no ruijinen*, Dobutsusha, Japan.

Kano, T., 1992, *The Last Ape: Pygmy Chimpanzee Behavior and Ecology*, Stanford, CA: Standord University Press.

Kortlandt, A., 1976, Letters: Statements on pygmy chimpanzees, *Primate Newsletters*, 15(1): 15-17.

Kortlandt, A., 1995, A survey of the geographical range, habitats and conservation of the pygmy chimpanzee, *Primate Conservation* 16: 21-36.

Kortlandt, A., 1996, The conservation status of *Pan paniscus*, *African Primates* 2(2): 79-80.

Lee, P., Thornback, J., Bennett, E.L., 1988, *Threatened Primates of Africa: The IUCN Red Data Book*, IUCN, Gland and Cambridge.

Malenky, R.K., Thompson-Handler, N., and Susman, R.L., 1989, Conservation status of *Pan paniscus*. Pp. 362-368 in: (Eds. P.G. Heltne and L.A. Marquardt), *Understanding Chimpanzees*, Cambridge, MA: Harvard University Press.

Meder, A., Herman, P., and Bresch, C., 1988, *Pan paniscus* in Salonga National Park, *Primate Conservation* 9: 110-111.

Muchaal, P. and Ngandjui, G., 1995, Secteur ouest de la reserve de faune de Dja (Cameroun): Evaluation de l'empact de la chasse villageoise sur les populations animales et propositions d'amenagement en vue d'une exploitation rationelle, *ECOFAC/MEF*, Yaounde, Cameroon.

Noss, A.J., 1995, *Duikers, Cables and Nets: A Cultural Ecology of Hunting in a Central African Forest*. Ph.D. dissertation, University of Florida, Gainseville.

Susman, R.L., 1984, *The Pygmy Chimpanzee: Evolutionary Biology and Behavior*, New York: Plenum.

Thompson, J., 1998, Conservation of *Pan paniscus* at the southern most research site, *Pan Africa News* 5(2): 22-24.

Thompson, J., 1999, Logging the Lukuru: Update on the conservation of *Pan paniscus*, *Pan African News* 6(2): 13-15.

Thompson, J., this volume.

Thompson-Handler, N., Malenky, R.K., and Reinartz, G.E., 1995, *Action Plan for Pan paniscus: Report on Free-ranging Populations and Proposals for their Preservation*, Milwaukee, Wisconsin: Zoological Society of Milwaukee County.

USAID/DRC, 2000, *USAID/DRC's Environment Program: a History and Profile*, Kinshasa, Democratic Republic of Congo.

van den Audenaerde, D.F.E., 1984, The Tervuren Museum and the pygmy chimpanzee. Pp. 3-11 in: (Ed. Susman, R.L.), *The Pygmy Chimpanzee: Evolutionary Biology and Behavior*, New York: Plenum Press.

Van der Kerken, 1944, L'Ethnie Mongo. Verhand.in-8°, K.B.K.I., Sectie voor fMorele en Politieke Wetenschappen, XII, 1 and 2. Brussels.

Van Elsacker, L., Dupain, J., and Van Krunkelsven, E., 1997, Au nom de nos ancêtres, Royal Zoological Society Antwerp (Belgium), *Zoo Magazine* 62(3): 18-23.

Van Elsacker, L., Dupain, J., Van Krunkelsven, E., Verheyen, R.F., Vervaecke, H., Walraven, V., 1996, Missing links: Setting trends for future bonobo research. Workshop at the XVIth congress of the International Primatological Society and XIXth conference of the American Society of Primatologists, August 11-16, Madison, Wisconsin, U.S.A.: Abstract No. 109.

Wilkie, D., Sidle, J., and Boundzanga, G., 1992, Mechanised logging, market hunting, and a bank loan in Congo, *Conservation Biology* 6(4): 570-580.

Wilkie, D., Curran, B., Tshombe, R., and Morelli, G., 1998, Managing bushmeat hunting in Okapi Wildlife Reserve, Democratic Republic of Congo, *Oryx* 32(2): 131-143.

Wolfire, D., Brunner, J., and Sizer, N., 1998, *Forests and the Democratic Republic of Congo: Opportunity in a Time of Crisis*, World Resource Institute, Washington, DC: Papyrus Design and Marketing.

Chapter 5

THE STATUS OF BONOBOS IN THEIR SOUTHERNMOST GEOGRAPHIC RANGE

J.A. Myers Thompson
Director, Lukuru Wildlife Research Project, P.O. Box 5064, Snowmass Village, CO 81615-5064

1.   INTRODUCTION

The Lukuru Wildlife Research Project (LWRP) focuses on scientific research, conservation, and education specific to bonobo populations (*Pan paniscus*) within the 23,908 km^2 Administrative Zone Dekese, Province Kasai Occidental, Democratic Republic of Congo, Africa.   The name "Lukuru" was derived by combining the abbreviated names of the two major navigable water routes within this zone, the Lukenie and Sankuru Rivers. The World Conservation Union *Action Plan for African Primate Conservation* (IUCN, 1996; see also Oates, 1986) identified this area as a priority for ongoing study and conservation.   The hilly terrain within the project area consists of irregular forest and grassland mosaic habitat.   The southern limit of bonobo distribution is defined by the Sankuru River at the southern boundary of this area.   The project area is inhabited by four human ethnic groups, the NDengese, the IKolombe, the IYalima, and the ISolu, all of whom have traditionally observed taboos against the consumption of bonobos.

2.   BONOBO HABITAT AND ECOLOGICAL
          ADAPTATION

Fieldwork conducted intermittently between August 1992 through June 1998 confirmed the existence of large blocks of perennially dry grassland utilized by the resident bonobo communities (for more details see Thompson, 1997). The bonobos' distribution in this habitat challenges the established view that these are arboreal apes specialized for lowland, moist forest environments. Although undisturbed climax forest is the predominant vegetation cover at this site and the bonobos clearly require access to forest, within the project region they occupy and use a drier and more open habitat than was previously known. The project area is characterized by two seasons: a rainy season from September through April, and a dry season from May through August. Compared to other bonobo study sites, within the Lukuru there is a wider range of daily and annual variation in temperature. The vegetation zone is in transition away from moist forest. Data from other bonobo study sites indicate perennially wet habitat with bimodal distribution of greater or lesser peaks of rainfall throughout the year and less variable daily and seasonal temperatures.

Grassland fruits (specifically *Landolphia lanceolata*, *Annona senegalensis*, and *Anisophyllea quangensis*) are consumed by the unprovisioned bonobos under study. Bonobos have not been observed eating these fruits at other sites. These findings suggest that bonobos were able to forage on grassland fruits in a mosaic ancestral habitat. Chimpanzees (*Pan troglodytes*) living along the northern periphery of the African tropical forest belt reportedly consume the fruit of *Annona senegalensis*, a savanna shrub (Goodall, 1968). Bonobos stride bipedally between these shrubs using their hands to carry fruits. *Anisophyllea quangensis* is a geoxylic suffrutice characteristic of the vegetation zone in transition from the African tropical forest belt to the great southern savanna region. Geoxylic suffrutices are confined to regions with marked seasonal distribution of rainfall. Some forest trees are replaced by suffrutices where seasonal waterlogging followed by seasonal drying out of sandy soil is prevalent. They share a common ancestor with large climax trees and responded to paleoclimatic extremes, while continuing to grow in unfavorable sandy soil conditions, by evolving into a massive subterranean structure. The patchy terrestrial appearance of this plant community is representative of the ancestral crown surface. While the primitive bonobo continued to consume this fruit, its distribution followed the oscillation of forest and grassland vegetation. Thus, the bonobo ancestor may have evolved within a broader, more southern geographic range.

3.   CONSERVATION ISSUES

During fieldwork between 1992 through 1998, visible amounts of climax forest were lost due to cutting for agricultural gardens and the demand for domestic wood usage by the ever-increasing human population. Legal changes effective May 1997 under the Kabila regime, and human demographic movement resulting from the eastern advance of those associated with the military sweep during the government takeover, resulted in two series of events.

First, a zone meeting was held on 13 December 1997, with all chiefs and officials from the entire region, to discuss hunting affairs. In 1937, a colonial decree prohibited hunting throughout the national territory. This was repealed in 1985 by national legislation that strove to be less restrictive in order to incorporate the needs of indigenous people. Hunting "seasons" were adopted at the discretion of the local authorities, and rules for the distribution of meat from particular species were established as the method of protection. Killing of "unprotected" species did not incur taxation or require distribution beyond the hunting party participants. Prior to the political changes during 1997, regional hunting of bonobos and Thollon's red colobus monkey (*Procolobus [badius] tholloni*) was prohibited by this law. However, the 1997 administration determined that only elephants, buffalo, hippopotamus, bongo, sitatunga, giant pangolin, leopard, yellow-backed duiker, bushbuck, and aardvark would continue to be "protected" at the regional level. Local officials decided that Thollon's colobus had become so rare that the species was no longer of consequence. Past, in some instances gender specific, taboos against ingestion of bonobo, black mangabey (*Lophocebus aterrimus*), and the mantled or Angolan black and white colobus monkey (*Colobus angolensis*) are being ignored by the younger generations. Fortunately, our focused education outreach campaign (December 1997 through May 1998) helped to officially restore the bonobo's protected status effective May 1998.

Second, the main commercial center between the Lukenie and Sankuru Rivers established a permanent market in 1998, the first in this administrative zone. Village leaders recruited BaTetela traders, from the ethnic group directly to the east, to organize the Yasa marketplace and trading base because of the BaTetela reputation for commercial expertise. The BaTetela consider bonobo meat a dietary delicacy, and they have exterminated the bonobos that once occupied eastern forests. The BaKuba ethnic group to the southeast developed a small-scale commercial route to transport dried bonobo meat to the market in Mweka, a three-day walk from Yasa. Two adult female bonobos were killed by BaKuba hunters in 1997. In past years, through taxation the regional government sanctioned killing bonobos as a source of cash revenue earned from the live animal trade and as official gifts for government dignitaries. I found, through monitoring the economics of local subsistence hunting, that there had been a disproportionate number of female bonobos killed to acquire their infants to support trade in live animals. Prior to 1998, the commercial trade in live bonobos was perhaps the most pervasive threat to the entire bonobo population. The demand for infant bonobos has also been reported from neighboring countries, such as the Republic of Congo.

During March and April 1998, my team members and I completed a species status survey and education campaign within the zone. This survey included a part of the southern portion of the Salonga National Park-South Block. Our survey was the first South Block bonobo project in the history of the Salonga National Park. Escorted by a unit of armed guards, we ascertained that the Libinja (a nationally organized group of notorious wildlife poachers) are no longer active in that sector of the Park. However, villages of IYalima people located throughout the Park's South Block continue to practice habitat modification and subsistence hunting within Park boundaries.

Accompanied by two members of our local project team, I surveyed a total area of 551 km and investigated 64 locations. Bonobos were present at 42% of the locations investigated. The survey area was divided by the Lukenie River, an impassable open water barrier that divides the bonobos inhabiting the study area into two subpopulations. Paleoclimatic data suggest that the bonobo subpopulation found south of the Lukenie River may have been isolated from the main population within the Congo Basin for as long as 8,000 years. North of the Lukenie River, 39 sites were investigated, and bonobos were found to be present at 38% of those locations. Within the southern sector of the survey, between the Lukenie and Sankuru Rivers, 25 sites were investigated, and bonobos were found to be present at 48% of those locations.

Conservation and education meetings were held throughout the zone. The women and young people of the region where we traveled do not speak French, Lingala (the Congolese trade language), or Kindengese (the language of the NDengese people, who administratively dominate the region), so it became important to share our information in their local language. Linguistically, the ISolu, Iyalima, and IKolombe come from the same mother language: Kikundu, the language of the NKundu people. They can therefore all understand each other. The NDengese have a different root language and cannot understand the other dialects within the region. All meetings were conducted by one of the Lukuru project team members in Kikolombe (translated into Lingala for me). Project stickers were distributed and eagerly accepted by the regional population. These stickers constitute a visible reminder of our message and remain displayed throughout the project zone.

Within the area between the Lukenie and Sankuru Rivers in Zone Dekese, the southern sector of the project area, there are 400-600 bonobos. Throughout this more defined area of 4,107 km, the project team has been able to identify at least five distinct, geographically separated bonobo communities. Additional bonobo communities inhabit the area but have not yet been geographically identified. Although the vegetation throughout the area appears to be homogenous, there are expanses of terrain between the communities where bonobos do not range. Recent efforts to observe bonobos have been concentrated within an area called Bososandja. Although I have rejected the suggestion from our local team to provision the Bososandja bonobo community, we constructed an observation blind alongside a perennial pool where we have been able to observe the bonobos drinking water and feeding on submerged vegetation. We have seen bonobos moving bipedally through the pooled water where their locomotion may be hampered by thick organic soil. Other observations occured as we encountered bonobos in the grassland blocks and forests of Bososandja.

In February 1998, we began a series of official meetings with the 21-member village council at Yasa to present details on the status of our project. I asked them to strongly enforce traditional taboos against consuming bonobo meat and to establish a specific tract of land where all wildlife would be protected. The designated tract of land is maintained with a strict prohibition against firearms, bows and arrows, nets, snare and steel-jawed traps, and fishing. Further, no trees may be cut for fields, fuel, building material, traditional medicines, pit-sawyer timber extraction, or any other use. The result of our efforts was the creation of a 3,400 ha wildlife sanctuary to maintain an intact ecosystem corresponding to the range of the Bososandja bonobo community. All wildlife will be protected within the sanctuary boundaries. The land for this sanctuary was acquired by local traditional law, and this transparent contract according to custom guarantees individual property rights to me as Project Director.

Although I have property rights and control the sanctuary, it is a community-based project that involves local participation through enforcement and education. By adopting civil authority, locally more suitable than a military one, we chose to avoid armed guards policing or defending the area against the local people. Due to increasing human population pressures, we will need to develop a long term solution that will attach an economic value to the bonobos other than for hunting and poaching and plan to encourage an alternative food source. Therefore, once security and stability return to this region, our site-specific, community-based conservation approach will facilitate sustainable, local, self-sufficient economic development without degrading the environment. As part of our conservation effort, the project has committed to sponsoring a regular domestic chicken inoculation and deworming program at local villages. The chicken population has a high incidence of morbidity and mortality. While on-site I frequently purchase chickens and eggs for personal consumption to stimulate local economics and actively promote the use of a domestically raised meat source. The project will also install a hand-pumped tubewell in the village of Yasa to access ground water for the community. The sealed well will need no filtration or cistern for water storage, thus reducing maintenance requirements. This is particularly relevant in this area because seasonal availability of water is a concern.

Prior to 1998, two pervasive threats throughout the bonobo's range were the loss of individuals as incidental catch in non-selective wire-snare traps and habitat modification from agricultural conversion. An emerging issue for the immediate future of our project area had been the threat of escalating habitat modification and human population growth through migration due to a potential lumber concession and the promise of a plywood manufacturing factory (Thompson, 1999; Thompson and Mvula, 1998). Similar to other sites (including Wamba, Lake Tumba, Yalosidi, Lilungu, and Salonga National Park) where bonobos have been or continue to be studied (but with the exception of the Lomako forest area, Badrian and Badrian, 1977; Dupain and Van Elsacker, this volume; Malenky, 1990; Thompson-Handler, 1990; Thompson-Handler, et al., 1995), the only timber extraction activity within the administrative zone corresponding to the project area and neighboring areas is small scale pit-sawing. To date throughout most of the bonobo's range there has been no active timber extraction or opening of the forest for roadways. Due to the absence of an overland infrastructure that would provide transport routes for commercial enterprise, only the forest shouldering navigable access waterways within the DRC are exposed to potential timber company activities. Within the bonobo's range, this is limited to the SIFORCO (previously known as SIFORZAL, representing the German company DANZER) concession corresponding to the Lomako Forest. The potential problem of timber company activity must be addressed at the administrative level. In March 1998 I met with the project director and international consultant of a group called "NICE GOODWILL SDN. BHD.," a subsidiary of ECHO ENTERPRISE SDN. BHD. based in Sarawak, Malaysia. They were on site conducting an initial reconnaissance of the forest hardwoods and logistics of the area. The forestry project personnel agreed to work cooperatively with us and to keep us informed of their progress. In follow-up communications, the group's Project Director of Forestry officially informed me of their decision to abandon their timber extraction project in the Lukuru project area effective October 1998.

ACKNOWLEDGMENTS

Since 1992, work for the Lukuru Wildlife Research Project has been funded in part by the Lukuru Wildlife Research Foundation; Columbus Zoo, Ohio; the National Geographic Society; the Foundation for Wildlife Conservation Incorporated; Primate Conservation, Incorporated; the Denver Zoo, Colorado; Flora and Fauna International; the Bonobo Protection Fund; the Royal Zoological Society of Antwerp; and the Twycross Zoo East Midland Zoological Society of Great Britain. I extend my appreciation to these groups.

REFERENCES

Badrian, A.J. and Badrian, N.L., 1977, Pygmy chimpanzees, *Oryx* 13: 463-468.

Dupain, J. and Van Elsacker, L., this volume.

Goodall, J., 1968, The behaviour of free-living chimpanzees in the Gombe Stream Reserve, *Animal Behaviour Monographs* 1(3): 161-311.

IUCN, 1996, *African Primates. Status Survey and Conservation Action Plan*, Revised Edition. IUCN, Gland, Switzerland.

Malenky, R.K., 1990, *Ecological Factors Affecting Food Choice and Social Organization in Pan paniscus*. Ph.D. dissertation, State University of New York at Stony Brook.

Oates, J.F., 1986, *IUCN/SSC Primate Specialist Group: Action Plan for African Primate Conservation 1986-1990*. Cambridge.

Thompson, J.A.M., 1997, *The History, Taxonomy and Ecology of the Bonobo (Pan paniscus Schwarz, 1929) with a First Description of a Wild Population Living in a Forest/Savanna Mosaic Habitat*. Ph.D. dissertation, The University of Oxford.

Thompson, J., 1999, Logging the Lukuru: Update on the conservation of *Pan paniscus*, *Pan Africa Newsletter* (2): 13-15.

Thompson-Handler, N.E., 1990, *The Pygmy Chimpanzee: Sociosexual Behavior, Reproductive Biology and Life History Patterns*. Ph.D. dissertation, Yale University.

Thompson, J. and Mvula, L.B.T., 1998, Conservation of *Pan paniscus* at the southern most research site, *Pan Africa Newsletter* 5(2): 22-24.

Thompson-Handler, N., Malenky, R.K., and Reinartz, G.E. (Eds.), 1995, *Action Plan for Pan paniscus: Report on Free-Ranging Populations and Proposals for their Preservation*. Milwaukee, Wisconsin: Zoological Society of Milwaukee County.

Chapter 6

CURRENT SITUATION OF BONOBOS IN THE LUO RESERVE, EQUATEUR, DEMOCRATIC REPUBLIC OF CONGO

C. Hashimoto[1] and T. Furuichi[2]

[1]*Primate Research Institute, Kyoto University, Kanrin, Inuyama, Aichi, 484-8506 Japan*

[2]*Laboratory of Biology, Meijigakuin University, Kamikurata, Totsuka, Yokohama, 244-8539 Japan*

1. INTRODUCTION

The Luo Scientific Reserve was established by the Zairean government in 1992 (Figure 17). This reserve is divided into northern and southern areas by the Luo River. There are small settlements of Wamba-Yokose peoples in the north and of Ilongo people in the south. Although most human activities are not restricted in the reserve, there are prohibitions against hunting primates, against hunting other animals using shotguns, and against planting new fields in the primary forest.

Ever since Kano started studying bonobos at Wamba in 1973, most research has been conducted in the northern area of the reserve (Kano, 1992; Kano, *et al.*, 1996). One unit group called "E group", which split into E1 and E2 groups in 1983, has been intensively studied, with occasional artificial feeding (Furuichi, 1987; Kano, 1982).

Although these studies have revealed many aspects of the ecology, social structure, and behavior of wild bonobos, there have been questions about how artificial feeding might influence study results. Researchers studying in the Luo Reserve are convinced that these impacts are minimized because of the small amount and limited period of artificial feeding in each year

(Furuichi, *et al.*, 1998). However, it is important to study unprovisioned groups of bonobos and to compare data collected with results from the provisioned study groups in the north.

The southern area of the Luo Reserve is a promising location for such a comparative study. Although the two regions of the reserve are separated by a wide river, they are included in an expanse of undisturbed forest. There is a smaller human population in the south, and the killing of bonobos is traditionally prohibited in both regions. Although only preliminary surveys have been conducted in the south, local field assistants confirmed the existence of several groups of bonobos.

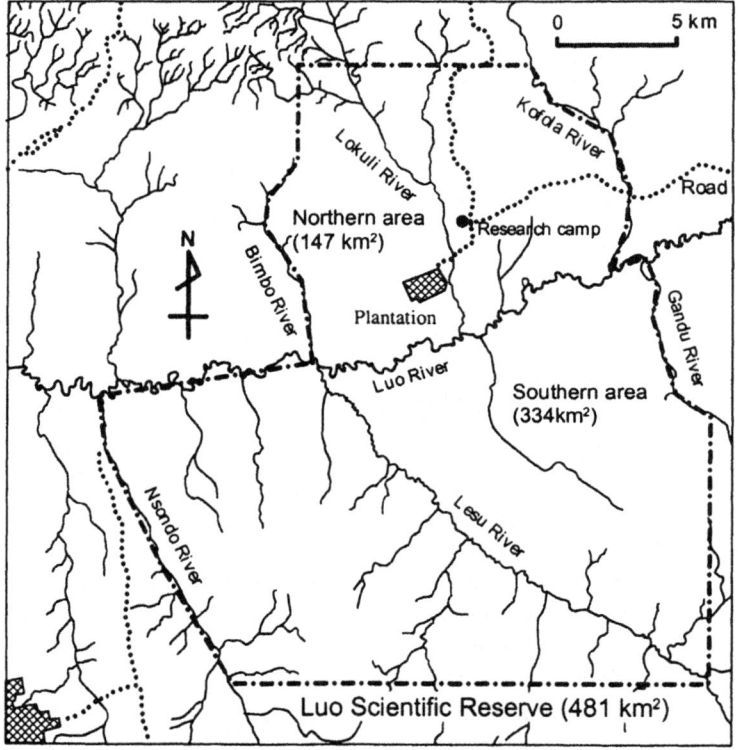

Figure 17. Map of the Luo Scientific Reserve

We report here the results of bonobo censuses conducted in both the north and south. For the north, we also estimated population density using known group size and ranging area of the main study groups. By comparing these results, we estimated population densities of bonobos in both areas of the Luo Reserve.

2.   MATERIALS AND METHODS

Population estimates by nest counts were conducted in both the northern and southern areas of the Luo Scientific Reserve, Equateur, Democratic Republic of Congo (DRC). To search for bonobo nests, we walked census routes with two local assistants at a speed of less than 1 km per hour. The total length of census routes in the north was 83 km and that in the south was 23 km. When we found a nest along the census route, we recorded the perpendicular distance from the census route to the nest. We ignored those nests that were not visible from census routes.

Based on the methods described by Ghiglieri (1984) and Buckland (1985), we used the following equation to estimate population density, *D* (individuals per km^2):

$$D = \frac{n}{2 \; L \; \mu} \times \frac{1}{d \; l \; b} \, ,$$

where *n* is the number of nests found along census routes, *L* is the total length of census routes, *d* is the number of nests built per individual per day, *l* is the mean life span of nests and *b* is the proportion of nest-building individuals. μ is given by:

$$\mu = \int_0^\infty g(x)dx \, ,$$

where $g(x)$ represents a detection function of the nests (Buckland, 1985).

We assumed that bonobos built one nest per day (see also Furuichi, *et al.*, 1997; Ghiglieri, 1984; Hashimoto, 1995; Tutin and Fernandez, 1984). We used 110.8 days for the mean life span of nests, which was calculated by Ghiglieri (1984) from his study of chimpanzees in the Kibale Forest, Uganda. We assumed that the proportion of nest-building individuals was 0.67, because about two-thirds of the bonobos of E1 group built nests.

Population density in the north was estimated from the known population size and ranging area. We used data on total population size and total ranging area of two neighboring groups, E1 and E2, collected in 1996. Because we could not follow all ranging movements of the study groups, especially when they ranged very far or in a swamp forest, the observed ranging alone greatly underestimates their actual ranging areas. To avoid biases, we assumed an approximate ranging area from direct observation and other information such as locations of vocalizations, direction of movement of the groups when they were lost or found, general tendency of movement of the groups, and eyewitness information provided by research assistants and local people.

We divided an area covering the home range of the two study groups into squares of 500 m. A square that was used by both study groups and other neighboring groups was counted as 0.5 square in our estimation of the size of each study group's ranging area.

3.   RESULTS

We counted 170 nests in the north and 25 nests in the south. Figure 18 shows the number of nests observed at each perpendicular distance from the census routes. This frequency distribution was approximated by a function of a half-normal distribution: $f(x) = a \exp(-bx^2)$ (χ^2=4989, df=6, p<0.001). Because a is a scale parameter, this approximation gave a detection function of $g(x) = \exp(-0.0012x^2)$.

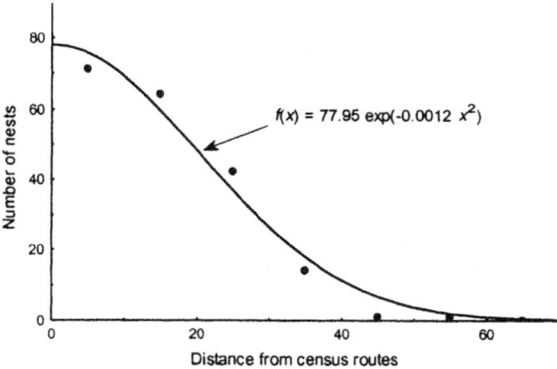

Figure 18. Frequency distribution of nests found from census routes

Using this detection function, we estimated the population density of bonobos in both the north and south. Density was 0.54 individuals per km^2 in the north and 0.28 individuals per km^2 in the south. Population density for the entire Luo Reserve was 0.49 individuals per km^2, which we obtained by pooling the number of observed nests and the length of census routes of both areas.

For the north, we estimated population density and ranging area. Population size in 1996 was 20 individuals for E1 group and 35 individuals for E2 group (Furuichi, *et al.*, 1998; Hashimoto, *et al.*, 1998), giving a total population size of 55 individuals. The total ranging area of these two groups was estimated to be 76 km². From these values, we calculated density in the north to be 0.72 individuals per km².

Table 6. Estimated number of bonobos in the Lou Reserve

	Area (km²) (1)	Denisty estimated from nest counts (ind./km²) (2)	Estimated number of individuals (1)x(2)	Estimated from group size and ranging area (ind./km²) (3)	Estimated number of individuals (1)x(3)
Northern area	147	0.54	79	0.72	106
Southern area	334	0.28	94	0.37*	125
Total	481		173		231

* Estimated from the density in the northern area, using the proportion of density estimated from nest counts.

Table 6 shows the estimated number of bonobos in the Luo Reserve. We made two estimations because there was considerable difference between the population density estimated from nest counts and that estimated from group size and ranging area. From these results, we infer that there were 173 to 231 bonobos in the Luo Reserve in 1996.

4.   DISCUSSION

The population density in 1996 calculated by us is lower than that estimated in 1982 by Kano and Mulavwa (1.7 individuals per km²). Furthermore, our density is lower than that reported for the Lomako Forest (about 2 individuals per km², Dupain and Van Elsacker, this volume). As Furuichi, *et al.* (1998) noted, population size of E1 group has been decreasing since 1991, and some members were probably killed by poachers. However, hunting pressure did not seem high enough to explain our low 1996 density. Further study is needed to specify the factors influencing population density and its change in the Luo Reserve.

In this study, we estimated population density of bonobos both from nest counts and from known group size and ranging area. In the north, the estimated density by nest counts was 25% lower than that calculated from group size and ranging area. There are several possible reasons for this difference.

- In this study, we used the mean life span of nests calculated for chimpanzees in the Kibale forest. However, life span of nests could be different in different forest types and for bonobos versus chimpanzees. We need to know the mean life span of bonobo nests at Wamba to yield a more accurate estimate of this parameter.

- Following many other studies based on nest counts, we tentatively assumed that one bonobo made one nest per day (Furuichi, *et al.*, 1997; Ghiglieri, 1984; Hashimoto, 1995; Tutin and Fernandez, 1984). However, Plumptre and Reynolds (1997) pointed out that this is not an appropriate assumption. They studied the number of day nests and the frequency of the reuse of old nests for chimpanzees in the Budongo Forest.

- They found that one chimpanzee built 1.09 nests per day. In the Luo Reserve, bonobos tended to frequently use several sleeping sites (Kano, 1992). If nest reuse is common, bonobos might make less than one nest per day. If so, the present analysis by nest count might underestimate the true bonobo population density.

This study showed that the population density of bonobos in the southern area of the Luo Reserve was about half of that in the northern area. Because of lower human density, hunting pressure on bonobos was expected to be lower in the south than in the north. Stronger hunting pressure cannot, therefore, explain the lower density of bonobos in the south.

The difference in vegetation types may account for the difference in the population densities of the two areas. A synthesized picture of Landsat data (path 178, row 60, 14 January 1991) suggests that the type of undisturbed forest is different between the northern area and the west side of the Lesu River in the southern area, where this study was carried out. The vegetation census of trees and herbs in the south also suggests differences in vegetation in north and south (Hashimoto, unpublished data). For example, a favorite herbaceous food of bonobos, *Megaphrynium macrostachum,* is abundant in the north but rare in the south. Because herbaceous food plays an important role in bonobo nutrition (Itani, *et al.*, 1994; Kano, 1992; Malenky and Stiles, 1991; Malenky, *et al.*, 1994; Wrangham, 1986), low density of this food source might be one reason for the lower density of bonobos in the south. Some tree species that produce favorite fruit foods, such as *Uapaca guineensis*, *Musanga cecropioides*, and *Pancovia laurentii,* are common in the north, but are rare in the south.

The lower density in the south might also result from the smaller area that we sampled, which may mean that we missed preferred sleeping sites. The tree species *Scorodiphloeus zenki* is abundant in the north but is less common in the south. Bonobos use this species as a food source and as a preferred nesting site (Kano, 1992). Because bonobos tend to use specific trees to make nests (Kano, 1992), the scarcity of this tree species might account for the low density of nests sampled in the south. The area we sampled in the south included dry primary forest, swamp primary forest, and secondary forest, but satellite (Landsat) imagery shows that there are other vegetation types in the vicinity. If bonobos prefer to sleep in other vegetation types, the population density estimated from nest counts in this study could be lower than the actual density.

We often found the food remains of *Haumania liebrechtsiana* and heard vocalizations of bonobos in the south, and so we had the impression that the population density in the south might be higher than in the north. Additional censuses that cover the areas of different vegetation types in the south are needed to clarify these issues.

ACKNOWLEDGMENTS

We thank Prof. T. Kano, Prof. T. Nishida, and members of the Primate Research Institute and Laboratory of Human Evolution, Kyoto University for their valuable help and advice for this study. We appreciate the cooperation of the Research Center for Ecology and Forestry of Congo (C.R.E.F.) and the people of Groupement Wamba-Yokose and Ilongo, Equateur, DRC. This study was supported by grants under the Monbusho International Scientific Research Program awarded to T. Kano.

REFERENCES

Buckland, S.T., 1985, Perpendicular distance models for line transect sampling, *Biometrics* 41: 177-195.

Dupain, J., and Van Elsacker, L., this volume.

Furuichi, T., 1987, Sexual swelling, receiptivity, and grouping of wild pygmy chimpanzee females at Wamba, Zaire, *Primates* 28: 309-318.

Furuichi, T., Inagaki, H., and Angoue-Ovono, S., 1997, Population density of chimpanzees and gorillas in the Petit Loango Reserve, Gabon: Employing a new method to distinguish between nests of the two species, *Int. J. Primatol.* 18: 1029-1046.

Furuichi, T., Idani, G., Ihobe, H., Kuroda, S., Kitamura, K., Mori, A., Enomoto, T., Okayasu, N., Hashimoto, C., and Kano, T., 1998, Population dynamics of wild bonobos (*Pan paniscus*) at Wamba, *Int. J. Primatol.* 19: 1029-1043.

Ghiglieri, M.P., 1984, *The Chimpanzees of Kibale Forest*, New York, Columbia University Press.

Hashimoto, C., 1995, Population census of the chimpanzees in the Kalinzu Forest, Uganda: Comparison between methods with nest counts, *Primates* 36: 477-488.

Hashimoto, C., Tashiro, Y., Kimura, D., Enomoto, T., Ingmanson, E.J., Idani, G, and Furuichi, T., 1998, Habitat use and ranging of wild bonobos (*Pan paniscus*) at Wamba, *Int. J. Primatol.* 19: 1045-1060.

Itani, G., Kuroda, S., Kano, T., and Asato, R., 1994, Flora and vegetation of Wamba forest, central Zaire with reference to bonobo (*Pan paniscus*) foods, *Tropics* 3: 309-332.

Kano, T., 1982, The social group of pygmy chimpanzees (*Pan paniscus*) of Wamba, *Primates* 23: 171-188.

Kano, T., 1992, *The Last Ape: Pygmy Chimpanzee Behavior and Ecology*, Stanford, CA: Stanford University Press.

Kano, T. and Mulavwa, M., 1984, Feeding ecology of the pygmy chimpanzees (*Pan paniscus*) of Wamba. Pp. 233-274 in: (Ed. Susman, R.L.), *The Pygmy Chimpanzee*, New York: Plenum Press.

Kano, T., Lingomo-Bongoli, Idani, G., and Hashimoto, C., 1996, The Challenge of Wamba. Pp. 68-74 in: (Ed. Cavalier, P.), *Great Ape Project, Ecta and Animali, 96/8*, Milano.

Malenky, R.K. and Stiles, E.W., 1991, Distribution of terrestrial herbaceous vegetation and its consumption by *Pan paniscus* in the Lomako Forest, Zaire, *Am. J. Primatol.* 23: 153-169.

Malenky, R.K., Kuroda, S., Vineberg, E.O., and Wrangham, R.W., 1994, The significance of terrestrial herbaceous foods for bonobos, chimpanzees, and gorillas. Pp. 59-75 in: (Eds. R.W. Wrangham, W.C. McGrew, F.B.M. de Waal, and P.G. Heltne), *Chimpanzee Cultures*, Cambridge, MA: Harvard University Press.

Plumptre, A.J. and Reynolds, V., 1997, Nesting behavior of chimpanzees: Implications for censuses, *Int. J. Primatol.* 18: 475-485.

Tutin, C.E.G. and Fernandez, M., 1984, Nationwide census of gorilla (*Gorilla g. gorilla*) and chimpanzee (*Pan t. troglodytes*) populations in Gabon, *Amer. J. Primatol.* 6: 313-336.

Wrangham, R.W., 1986, Ecology and social relationships in two species of chimpanzee. Pp. 352-378 in: (Eds. D.L. Rubenstein and R.W. Wrangham), *Ecological Aspects of Social Evolution*, Princeton: Princeton University Press.

SECTION THREE
CHIMPANZEES, THE BEST KNOWN APE
INTRODUCTION

Chimpanzees (*Pan troglodytes*) were among the first primates described by western philosophers (Goodall, 1986), and since Darwin's and Huxley's times, the species' importance as a close human relative has been emphasized (Zihlman, 1997). Perhaps no other nonhuman primate species has been the subject of such intense scientific scrutiny in both captive and field settings. Discoveries of chimpanzees' hunting, tool-making, language, and cultural behaviors have further fostered comparisons to modern humans and our ancestors. Rose (this volume) notes that western populations have developed a strong affinity and even a sense of kinship with this ape, primarily through such comparisons. However, in opening Section Three, Sarah Boysen and Tom Butynski provide scant evidence to show that this western sensibility has had a positive impact on wild *or* captive chimpanzees.

While descriptions of the chimpanzee's behavioral complexity continue to fascinate us, the species is often the preferred biomedical model, is highly sought after as a pet, and is subject to the same environmental pressures that threaten other African species. In the 2000 World Conservation Union (IUCN) *Red List of Threatened Species*, experts recognize one chimpanzee species with four subspecies. The species as a whole and each of its subspecies are classified as endangered, with an expectation that at least half the remaining population will be lost in the next three (chimpanzee) generations. Meanwhile, chimpanzees are bred for scientific research, the pet trade, and the entertainment industry with little forethought for the adult animals' needs throughout a 40-year or more life span. Boysen and Butynski describe an apparent paradox of too few wild chimpanzees and too many captive ones kept under inadequate and oftentimes inhumane conditions. They make an impassioned plea for improved care of and heightened concern for this species, which, along with the bonobo, is our closest living relative.

The discovery that chimpanzees are hunters has had a major impact on interpretations of the human fossil record and the social and nutritional significance of hunted meat in the human diet (Stanford, 1999, 2001). Since the first observations of chimpanzee predation, researchers at several long term sites (primarily Gombe, Ngogo, Bossou, and Mahale) have noted variation in hunting techniques, prey items selected, and predation success rates. These differences reflect in part varying chimpanzee traditions, ecologies, and demography, but different data collection techniques make intersite comparisons difficult. Kazuhiko Hosaka, Toshisada Nishida, Miya Hamai, Akiko Matsumoto-Oda, and Shigeo Uehara present a longitudinal analysis of chimpanzee predation at Mahale. Through comparisons with data collected at Ngogo, Gombe, and Bossou, the authors describe common features of chimpanzee predation: the focus on forest dwelling monkeys, the existence of group hunting, and seasonal fluctuation in predation on mammals. However, hunting methods differ between sites, as does prey selected by chimpanzee predators. Hosaka and his colleagues attempt to account for this variation while controlling for differences in data collection techniques and study durations. Possible explanations include the composition of hunting parties, the status of individuals comprising the hunting party, the availability of prey items versus other preferred foods, and local ecology.

Studies of wild chimpanzees hint at the complex intellectual processes that underpin chimpanzee behavior, but these processes are difficult to verify. In laboratory or other captive settings, questions of cognition can be considered in more detail, and comparisons between humans and nonhumans help us to comprehend the minds of our closest relatives. Sarah Boysen and Valerie Kuhlmeier close Section Three with a presentation of chimpanzees' abilities to "map the physical world onto multiple representational systems". They use scale models, photographs, and videotapes to test apes' understandings of these representations of actual events. Tests of human children aged 2.5 and 3 years show that younger children do not understand the models as a symbol or representation of an actual room, but older children do. Boysen and Kuhlmeier tested whether chimpanzees are capable of forming such an orientation, and find that this is, indeed, possible. Adult female chimpanzees, using photographs, videos, or scale models, understood that what was depicted in the model was analogous to events that occurred in the actual room. However, male subjects were unable to do so, indicating intraspecies variation in the ability to perform this task. The amazing cognitive complexity of chimpanzees as illustrated by Hosaka and his colleagues and Boysen and Kuhlmeier makes the slaughter of wild chimpanzees even more tragic.

REFERENCES

Goodall, J., 1986, *The Chimpanzees of Gombe: Patterns of Behavior*, Cambridge, MA: Harvard University Press.

IUCN, 2000, *Red List of Threatened Species*. Eletronic document, http://www.iucn.org, accessed January 2001.

Rose, A.L., this volume.

Stanford, C.B., 1999, *The Hunting Apes: Meat Eating and the Origins of Human Behavior*, Princeton, NJ: Princeton University Press.

Stanford, C.B., 2001, The ape's gift: Meat-eating, meat-sharing, and human evolution. Pp. 97-117 in: (Ed. F.B.M de Waal), *Tree of Origin: What Primate Behavior Can Tell Us about Human Social Evolution*, Cambridge, MA: Harvard University Press.

Zihlman, A., 1997, African apes. Pp. 17-23 in: (Ed. F Spencer), *History of Physical Anthropology, Volume 1 (A-L)*, New York: Garland Publishing, Inc.

Chapter 7

PAN IN PANDEMONIUM

S.T. Boysen[1] and T. Butynski[2]

[1]*Comparative Cognition Project, Dept. of Psychology, The Ohio State University, Columbus, OH 43210-1222; Living Links Center for Human and Ape Evolution and Behavior, Emory University, Atlanta, GA; Great Ape Aging Project, Diagnon Corporation, Rockville, MD*

[2]*Zoo Atlanta, Africa Biodiversity Conservation Program, P.O. Box 24434, Nairobi, Kenya*

1.   INTRODUCTION

The chimpanzee (*Pan troglodytes*), like all of the great apes, is currently facing dire circumstances. While some may quibble as to its status relative to humankind compared with its close cousin, the bonobo, or relative to gorillas or orangutans, there is little question that the chimpanzee species is losing its "common" status all too soon. The chimpanzee, again like all the apes, is now in a precarious situation that might be characterized as too much, or perhaps too little, and likely too late. Devastation to wild chimpanzee populations throughout equatorial Africa, exploitation by some sectors of the research community, and potential threats to the long-term care of animals that were purposefully bred in captivity (either for biomedical research or the entertainment industry, including the pet trade by private breeders), require attention in all three directions simultaneously and immediately, a daunting task indeed.

These three avenues of destruction for the chimpanzees, including habitat destruction and predation by humans in the wild, use of chimpanzees for invasive studies in the West, and continued breeding for a kind of planned obsolescence for the entertainment business, all share features of too much, too little, and too late, in different ways. None bode well for the chimpanzee's future. This is not a welcome message, but nevertheless, it is the very real and pressing current situation for the species that shares a special status with the human species' past, present, and future.

2.   CHIMPANZEES IN THE WILD

Our present remarks are an outgrowth of invited talks and resulting discussions at recent meetings, spearheaded by Ben Beck (National Zoological Park/Smithsonian), organized to discuss the ethical frontiers for great apes in captive and field settings (see Beck, *et al.*, in press). Much of the information concerning the status of wild chimpanzees comes from extensive work and data summaries compiled by one of us (Butynski, in press), reports and data from Jane Goodall, and a recent population and habitat viability assessment report that focused on conservation efforts with wild chimpanzee populations in Uganda, (Edroma, *et al.*, 1997). The report by Edroma and his colleagues (1997) includes a wealth of important findings that were presented at a workshop held in Entebbe, Uganda in January 1997.

2.1   Distribution of Current Chimpanzee Populations

The current distribution of the chimpanzee (*Pan troglodytes*) is thought to encompass approximately 23 countries throughout equatorial Africa (IUCN *Red List of Threatened Species*, 2000). It should not be surprising, given the exponentially exploding human population throughout Africa, that the distribution of chimpanzees has become extremely fragmented over the past 20-30 years as agricultural and economic developments have rapidly encroached upon their habitat. Such fragmentation reduces opportunities for natural exchange of chimpanzee individuals, and consequently genetic material, among these populations. This isolation, in turn, threatens the stability of each of the subpopulations and greatly enhances the risks of local extinction. The burgeoning human population, which is declining nearly everywhere else in the world yet rising dramatically in Africa, represents the greatest single threat to wild chimpanzees. From the simple formula of too many people and too much demand for too few resources, the result is havoc for the chimpanzee through deforestation and loss of habitat (Figure 19).

Figure 19. Ascending deforestation and destruction of habitat typical in Africa.

In addition, the sale of bushmeat has now reached crisis proportions, and it is particularly critical in particular African countries, such as Cameroon. There are other natural and acquired sources of loss to chimpanzee populations, including disease or disasters such as fires or local wars. However, clearly the most significant pressure is forest destruction for logging and agricultural use. With logging expansion has come an even greater threat to wild populations besides habitat loss, since newly-created roads now allow hunters to access chimpanzee populations within what was once inaccessible forest. Such logging expansion has had a dramatic impact on the bushmeat trade, moving it far beyond subsistence operations into an opportunity to provide a food source for the logging camps (Dupain and Van Elsacker, this volume; Rose, this volume). Access to deeper chimpanzee habitat has helped create a highly desirable consumer product for the restaurants and palates of upper and middle class populations in central Africa.

2.2   The Greatest Threat: Habitat Loss

Forest habitats are being destroyed more rapidly in Africa than anywhere else in the world. Recent statistics (Butynski, in press) are not merely alarming, but reveal a catastrophic trend:

- Current estimates are that in the next 40 years, 70% of Africa's forests will be gone;
- In Burundi, Kenya, and Uganda, 95% of the forests will be gone in the next 40 years;
- Currently, $^1/_3$ to ½ of western chimpanzees (*Pan troglodytes verus*) live in the forests of the Ivory Coast, and all forest there will be gone within the coming decade;
- While there are regulations for deforestation practices in most African nations, enforcement is difficult. The result has been unmanaged and uncontrolled timber extraction.

Disease is also emerging as a problem for chimpanzees in Africa (Wallis and Lee, 1997). Transmission from humans to wild apes has received little attention, but apes are susceptible to parasites, viruses like *Ebola*, respiratory diseases such as colds and flu, malaria, polio, hepatitis, whooping cough, tuberculosis, rubella, chicken pox, and mumps, all of which can be deadly. Consequently, the incursion into chimpanzee habitat where previous human contact had been minimal, by hunters, miners, loggers, farmers, and eco-tourists (which brings porters, rangers, guides, visitors, and researchers into closer proximity to wild populations) has the potential for passing a wide range of diseases to these already vulnerable populations. Add the stress factors associated with forced habituation, death of a group member due to hunting, and other devastating events, all of which can contribute to compromising the immune systems of individuals or entire groups of chimpanzees, and the potential impact of new disease transmission can be deadly. For example, at Mahale, near a tourist site, 11 animals died recently, another individual and perhaps 11 others died from *Ebola* in a second incident, and this tragedy reduced the number of males in this important, long-term study from eight adults to a mere two. At Gombe Stream Reserve in Tanzania, the famous site of Goodall's longitudinal study of wild chimpanzees, there were several significant disease outbreaks between 1966-1990. In all cases, it has been proposed that the diseases were transmitted by people, including an early outbreak of polio that killed six animals and crippled six others, with three additional chimpanzees missing. In 1986, twelve chimpanzees at Gombe died from pneumonia, followed in 1996 by the death of seven animals from a respiratory virus. The most recent report was for 1997 during which there was an outbreak of sarcoptic mange, with some deaths and a number of chimpanzees losing all their hair, but surviving. This represents a total of 39 animals seriously affected or dead at the Gombe site alone over a very brief period of time.

2.3   Uganda: A Model Program for Chimpanzee Conservation

We now consider a specific program in Uganda that Richard Wrangham (in press) considers to be a conservation success story. Of course, problems persist for the chimpanzee, even in a country such as Uganda that does not permit the holding of chimpanzees as pets and where government officials confiscate chimpanzees at the airport and at border posts, with animals then transferred to the Entebbe Wildlife Education Center. Logging is also banned, and the Ugandan people are opposed to the exploitation of the forest and its primates. There is no bushmeat trade or eating of primates. Even with such strong governmental and conservation support, there can still be problems. Wrangham (in press) suggests a three-way approach to on-site problems, including:

1. trust funds to insure longevity of funding for conservation efforts;
2. research activity which can, in and of itself, serve a protective function;
3. tourism activity, to provide jobs and other economic and educational incentives for the local populations to have a portion of ownership in the conservation effort.

For the Ugandan chimpanzees, snares are still a hazard, although they are not intentionally set to capture chimpanzees. Instead, other small mammals are sought. In some cases, this hunting behavior has become a type of thrill seeking by young men. It is generally the case that more male chimpanzees are caught in snares than females, as males lead the group in foraging and movement. Wrangham (in press) estimates that 20% of the Kibale chimpanzees have snare damage in the form of lost or damaged limbs, and that the figure rises to 30% in the Budongo Forest. Chimpanzees also visit closer to park edges and consequently come into contact with local people. These issues, which are often site-specific, require immediate attention to protect both the chimpanzees and the people from harm. As elsewhere in Africa, disease problems exist in Uganda; for example, local farmers defecate in fields thereby creating a transmission route of bacteria and other pathogens to chimpanzees in the area.

At Kibale, Wrangham and his colleagues and students offer local seminars to talk about chimpanzees to the local people, with such educational outreach considered critical to building relationships between the local human populations and the chimpanzees to be protected in their area. This research team stresses the positive impact that tourism has had on the local community. Tourism thus can serve a very significant function for the long-term conservation effort for chimpanzees, as it helps to promote a type of personal investment in the animals that is admittedly difficult to replicate through other approaches and methods. Tourism has associated challenges that must be addressed, but in the case of the Ugandan program, it appears to be working with measurable success.

Edroma, *et al.*'s (1997) report at the Entebbe workshop included the following recommendations with respect to captive chimpanzee policies in Uganda. Captive chimpanzees should not be used for the following:

- Medical research;
- The entertainment industry;
- The pet trade;
- Should not be privately held;
- Young chimpanzees should not be exhibited at schools and fairs;
- Holding of chimpanzees should follow guidelines of international zoo regulations;
- Management of captive chimpanzees should be done under the guidance of a recognized management committee;
- Education of police, customs agents, and other officials should start as soon as possible and should be an on-going process;
- Only non-invasive studies should be allowed on captive chimpanzees, with the emphasis on research that will benefit their management.

The report also recommended the formation of an animal rescue unit to respond to confiscations, and in areas where such confiscations are numerous, information about potential fines and imprisonment should be relayed to local communities in concert with other educational efforts aimed at the conservation of chimpanzees and their habitat.

3.   THE CURRENT CRISIS FOR CAPTIVE CHIMPANZEES

Shifting to the plight of many captive chimpanzees in the United States, Jane Goodall urges us to think about the crisis from the perspective of the chimpanzees. This includes problems in zoos, the wild, the pet trade, the entertainment business, and the limited number of sanctuaries outside of Africa. For instance, there are currently only two sanctuaries caring for sizeable numbers of chimpanzees, including Primarily Primates in Texas and Wildlife Waystation in California. For captive chimpanzees the too much, too little, too late message has different connotations than for wild chimpanzees. The "too much" includes too much intentional breeding, particularly for biomedical research, although this issue has been recognized and most facilities holding large collections of chimpanzees, such as Southwest Foundation for Biomedical Research, Primate Foundation of Arizona, and the University of Texas M.D. Anderson Cancer Center at

Bastrop, Texas, have their animals on breeding moratorium. There continues to be, however, too much exploitation of chimpanzees through the media and entertainment business that nonetheless persist in promoting the chimpanzee as a way to poke fun at the human condition or otherwise demean or cast chimpanzees in a stupid or comical light. There also continues to be too much intense polarization between those who identify themselves with the animal rights, welfare, and humane community and with the biomedical research community, which created further divisiveness from the 1985 designation of the chimpanzee as endangered in the wild, but merely threatened in captivity, with the dual-listing lobbied vigorously by the biomedical community. The resulting intensity of the most extreme views within each of these different philosophical and pragmatic stances as to how captive chimpanzees should live and be managed almost certainly contributed to a worsening of the situation for the animals. Valuable time and resources have been squandered on both sides of the issue over the past several decades in retaliatory efforts, when consensus-building and communication should have been the absolute priority. This must change immediately. There is a place for passion and commitment for all of us to use our intelligence and resources to work in concert, to create more opportunities for the "win-win" resolution of these contentious issues, keeping foremost in our hearts and minds that chimpanzees lose every day that we delay our search for solutions by slipping back to rigid, all-or-nothing thinking.

The laboratory chimpanzee problem in the U.S. looms large. While great strides have been made in the last decade with the implementation of enrichment programs and social housing for chimpanzees, and there are fewer labs with singly-housed animals today, some single housing does still exist in cases where extreme types of medical protocols demand isolation. Such instances appear to be found in contract or commercial laboratories, where ease of cleaning, feeding, and animal management in general may override concerns for the psychological well being of the animals. Despite guidelines mandated by law and enforced through the U.S. Department of Agriculture (USDA), enforcement is still an issue, with too few inspectors, inspecting too seldom, and with almost no option available for absolute judgements of violations or confiscations. There are simply no facilities available to accommodate the chimpanzees that could potentially be seized by the USDA, and thus chimpanzees remain in the same deplorable conditions, sometimes for years or even decades. After an inspection with numerous violations, sometimes with repeated citations, USDA veterinary inspectors are often powerless to demand change of any substance in the animal's situation.

3.1   Too Many "Surplus" Chimpanzees; Too Few Sanctuaries

Another highly significant issue in the U.S. is the plight of the so-called "surplus" research chimpanzees, numbering perhaps 1,000 animals, that were purposely bred with the support of the government through the National Institutes of Health (NIH) Chimpanzee Breeding Program, which has successfully encouraged the care, breeding, and management of chimpanzees by funding five sites holding large populations of chimpanzees, to insure a ready supply of subjects for the biomedical community. The establishment of the NIH breeding program was a response to the CITES agreement in the 1970s that made further importation of wild chimpanzees illegal. That program was highly successful in producing a large chimpanzee population.

A more recent, widely-cited transfer of some 255 chimpanzees from a biomedical laboratory (LEMSIP), makes clear the difficulties in housing and maintaining this now burgeoning population of captive chimps. When the LEMSIP facility was closed several years ago, only a few chimpanzees from the collection made it to retirement-type sanctuaries, including 15 HIV-infected chimpanzees sent to a private sanctuary in Canada, seven young chimpanzees that went to another private sanctuary in Kentucky, and finally, a group of some 75+ chimpanzees sent to Wildlife Waystation outside Los Angeles, California. The rest of the colony was transferred to the ownership of the Coulston Foundation in New Mexico. Also leased to the Coulston Foundation were 143 chimpanzees that were once part of the U.S. Air Force's chimpanzee colony from which the original test animals for later manned space flights emerged. With an announced plan to divest itself of the Air Force colony, a dramatic media battle and competition for bids to procure the group of chimpanzees erupted and led these chimpanzees to become the "poster children" for the plight of all U.S. chimpanzees. Perhaps one good outcome has been the emergence of several organizations whose primary goal is the establishment of new chimpanzee sanctuaries. Indeed, if three to five such sanctuaries were constructed for 100 chimpanzees each, they would likely be filled within days. Eventually, following the bidding process used with the Air Force, a small group of 30 chimpanzees from the Coulston colony were released to Primarily Primates, a primate sanctuary in San Antonio, Texas, despite the fact that no funds were included in the Air Force divestiture plan. More recently, an animal right's organization, In Defense of Animals (IDA), won a legal battle against Coulston and was nominally awarded custody of some 300 chimpanzees, although an appeal is planned by the Coulston Foundation, and IDA will never likely take custody of the animals. On a more positive note, there has been an effort by a group of moderate individuals from the research and humane communities to draft legislation that would provide NIH funding for retirement and sanctuary facilities for government-owned, ex-research chimpanzees. A draft of this legislation has been recently submitted to Congress. However, even legislation that becomes law will not guarantee that money will be immediately, if ever, forthcoming, and so the issue of how to resolve the need for long-term care and management of powerful and demanding adult chimpanzees, including a proportion of whom have been intentionally injected with pathogens such as various hepatitis viruses or HIV, is an extremely difficult one.

3.2   Let Them Entertain Us; Let Them Make Us Smile

There is yet another persistent problem for captive chimpanzees in the U.S., and that is their exploitation for the entertainment industry and the pet trade. Surprisingly, much of the public is unaware that it is completely legal to purchase a chimpanzee (as well as many other nonhuman primates and other exotic animals) for private ownership in the U.S. There are only two or three states that do *not* permit citizens to have primates as pets. There are no exact records of acquisition of chimpanzees by private individuals nor records of the ultimate disposition of these animals, but it is likely the majority of chimpanzees that were first acquired as cute, cuddly infants end up in dire circumstances once they outgrow their physical and behavioral attractiveness as infants. Stories of chimpanzees escaping from insecure housing, including people's basements and garages, and ultimately being shot to death as the only means of controlling their rampage have been typical in the past two decades. There is no question that there are few accurate perceptions from either the lay public or the scientific community that realistically predict the eventual size, strength, and power of adult chimpanzees, particularly males. Even well meaning individuals who purchase chimpanzees are almost never prepared for the dramatic changes in infant chimpanzees as they move from infancy to adolescence. Again, the changes in motivation and physical size are more dramatic in males, who go through an exponential growth spurt as they approach puberty and subsequently require well-reinforced housing areas to safely maintain them.

Because young adult and adult chimpanzees are so resource-demanding in terms of proper veterinary care and safe and adequate housing, with appropriate space and furnishings for development and daily opportunity for species-typical behaviors, most private owners are forced to seek alternative housing for their animals for whom they often have deep affection. Unfortunately for the chimpanzees, suitable sites for retirement and/or sanctuary are outnumbered by the chimpanzees in need of such long-term care. The cost for construction of facilities that are appropriate for maintaining adult animals is formidable, as are the long-term costs for care and management (estimated to be about $20 per day per chimpanzee) with construction costs estimated at anywhere from $5-15 million, depending on design, location, regional labor costs, and materials. Unquestionably, long-term care for chimpanzees is a tremendous responsibility, and one that is not undertaken lightly. Collaborative efforts for the establishment of sanctuaries and retirement facilities for chimpanzees in need of long-term care must be encouraged among individuals from the research and lay communities, with emphasis on shared issues and concerns. A commitment to the chimpanzees must be the most critical priority.

There is a serious need for legislation at state and federal levels to better protect chimpanzees and other nonhuman primates from being sold as pets, and more funds for inspections and enforcement of current USDA regulations and guidelines for maintaining exotic animals by animal dealers, non-American Zoo Association approved zoos, and private individuals, to minimize situations such as the recent discovery of five adult chimpanzees, all in their mid-twenties, kept in a small concrete block room, with no permanent water source, minimal food, and no ventilation (Figures 20, 21, and 22). Since removed to a private sanctuary with access to clean and spacious indoor/outdoor quarters and proper veterinary attention, these chimpanzees are among the few lucky ones. We, as humans, remain critical to the stewardship of our ape kin. As they continue to suffer from our encroachment and short-sightedness, it brings us closer to our own destruction on an increasingly fragile planet with finite resources, including land, forests, and water, but, hopefully, not compassion.

Figure 20. Makeshift ex-jail cell used to house five adult chimpanzees.

Figure 21. Deplorable conditions under which these five chimpanzees were kept.

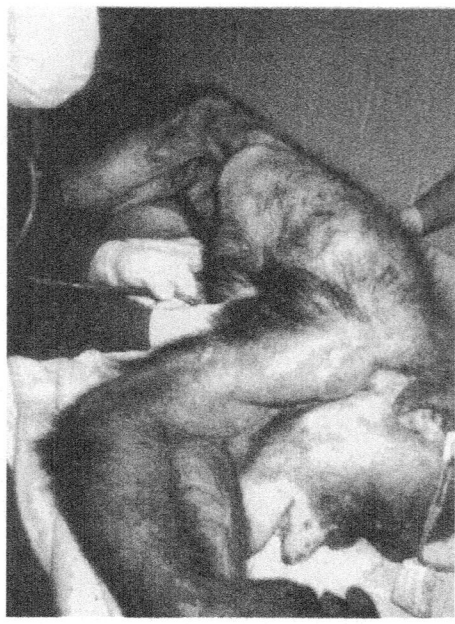

Figure 22. Adult female chimpanzee, weighing 29 kg, part of the group of five chimpanzees rescued from the conditions described above.

REFERENCES

Beck, B., Stoinski, T., and Maple, T. (Eds.), in press, *Great Apes and Humans at an Ethical Frontier*, Washington, DC: Smithsonian Press.

Butynski, T., in press, Africa's great apes: An overview of current taxonomy, distribution, conservation status, and threats. In: (Eds. B. Beck, T. Stoinski, and T. Maple), *Great Apes and Humans at an Ethical Frontier*, Washington, DC: Smithsonian Press.

Dupain, J. Van Elsacker, L., this volume.

Edroma, E., Rosen, N., and Miller, P., 1997, *Conserving the Chimpanzees of Uganda: Population and Habitat Viability Assessment for Pan troglodytes schweinfurthii*, Apple Valley, MN: SSC/IUCN Conservation Breeding Specialist Group.

IUCN, 2000, Red List of Threatened Species. Eletronic document, http://www.iucn.org, accessed January 2001.

Rose, A.L., this volume.

Wallis, J. and Lee, J.R., 1998, Primate conservation and health: II. Prevention of disease transmission, *Proceedings of a Symposium on Veterinarians in Wildlife Conservation.* 7th International Theriological Congress. World Association of Wildlife Veterinarians.

Wrangham, R., in press, Apes at the forest edge: Multiple problems, multiple solutions. In: (Eds. B. Beck, T. Stoinski, and T. Maple), *Great Apes and Humans at an Ethical Frontier*, Washington, DC: Smithsonian Press.

Chapter 8

PREDATION OF MAMMALS BY THE CHIMPANZEES OF THE MAHALE MOUNTAINS, TANZANIA

K. Hosaka[1], T. Nishida[2], M. Hamai[1], A. Matsumoto-Oda[2], and S. Uehara[1]
[1]*Primate Research Institute, Kyoto University, Inuyama, Aichi 484-8506 Japan*

[2]*Laboratory of Human Evolution Studies, Graduate School of Science, Kyoto University, Kyoto, 606-8502 Japan*

1.   INTRODUCTION

One of the most intriguing features of wild chimpanzees (*Pan troglodytes*) is their predatory behavior toward mammals on a regular basis, which has been shown by a number of studies covering all three subspecies (e.g., review: Uehara, 1997) since Goodall (1963) documented the first evidence that chimpanzees ate freshly-killed mammals. Four major long-term study sites in particular (Mahale in Tanzania: Uehara, *et al.*, 1992; Gombe in Tanzania: Stanford, 1998a; Taï in Ivory Coast: Boesch and Boesch, 1989; Ngogo in Uganda: Mitani and Watts, 1999) have contributed much to this issue, and several attempts have been made to show inter-population variation despite researchers' methodological differences at each site.

Several similarities have been proposed through such inter-population comparisons. First, Uehara (1997) pointed out that the tendency for chimpanzees of all subspecies to hunt forest monkeys, colobine species in particular, is ubiquitous provided that the prey monkeys are sympatric with the chimpanzee hunters. Interestingly, the proportion of red colobus (*Colobus badius*) in eaten prey carcasses has been reported at around 80% or

107

more at the three major long-term study sites (Gombe: 82%, n=429, Stanford, *et al.*, 1994a; Taï: 78%, n=81, Boesch and Boesch, 1989; Ngogo: 91%, n=128, Mitani and Watts, 1999) other than Mahale (53%, n=115, Uehara, *et al.*, 1992; but see below for recent data).

Second, it should be noted that chimpanzees of the four long-term study sites mainly exploit "group hunting," defined as a pattern where two or more hunters participate in chasing prey when they hunt arboreal monkeys (Boesch and Boesch, 1989; Mitani and Watts, 1999; Stanford, *et al.*, 1994a; Uehara, *et al.*, 1992), though its frequency and cooperation levels seem to vary among populations (Boesch, 1994a, b; Boesch and Boesch, 1989; Mitani and Watts, 1999). Group hunting of mammals rarely occurs in most other primates (e.g. review: Butynski, 1982), although in a few species some evidence of group hunting has been confirmed or suggested (savannah baboons: Hausfater, 1976; Strum, 1981; capuchins: Rose, 1997; mandrills: Kudo and Mitani, 1985). There is no evidence that other apes hunt in groups (e.g., bonobos: Ihobe, 1992; orangutans: Utami and van Hooff, 1997).

Finally, predation of mammals shows seasonal fluctuation at least at Gombe (Stanford, *et al.*, 1994a), Taï (Boesch and Boesch, 1989), and Mahale (Takahata, *et al.*, 1984). At both Gombe and Mahale, predation frequency was higher in the dry season than in the rainy season. At Taï , the hunting season seems to correspond to the period when Coula nuts are out of season. Mitani and Watts (1999) presented data showing an apparently seasonal pattern at Ngogo, but they did not investigate this topic extensively while analyzing the relationships between monthly hunting frequency and demographic factors.

Previous studies have also indicated some differences among populations. The first important difference is that the Mahale chimpanzees have been regarded as the most euryphagous predators in terms of prey selectivity (Uehara, 1997; but see updated information below), while the Taï chimpanzees are the most stenophagous (Boesch and Boesch, 1989). At Mahale, 16 mammal species have been confirmed as prey of chimpanzees (Uehara, *et al.*, 1992), contrasting with eight in Gombe (Goodall, 1986) and six in Taï (Boesch and Boesch, 1989). That is, the Taï chimpanzees eat primates only and neither hunt nor eat artiodactyls, though four species of duikers are encountered daily in the forest. On the other hand, the Mahale chimpanzees consume all sympatric artiodactyl species (Uehara, 1997; also see below).

Another notable difference appears to exist in hunting methods. Boesch and Boesch (1989) insisted that "cooperative hunting" is the remarkable feature of the Taï chimpanzees, whereas the Gombe and Mahale chimpanzees capture prey mostly in an uncoordinated manner. Group hunting of monkeys by the Gombe and Mahale chimpanzees was depicted as "simultaneous solitary hunting" (Boesch, 1994b; Nishida, 1981; but see updated information below). Boesch and Boesch (1989) emphasized ecological differences between the Taï and eastern Tanzania habitats as the key factor causing the differences between Taï and eastern Tanzania habitats. They hypothesized that the high canopy and high density of the Taï forest make solitary hunting costly in terms of energy, thus forcing chimpanzees to hunt monkeys cooperatively in contrast to the chimpanzees of the more open forests such as Gombe and Mahale. Boesch (1994a, b) collected comparable data both at Taï and at Gombe to strengthen his hypothesis. However, Mitani and Watts (1999) observed that the Ngogo chimpanzees did not cooperate in hunting monkeys, although the habitat structure seemed similar to Taï.

Predatory behavior of the Mahale chimpanzees has been documented since the initial stage of long-term fieldwork (Kawanaka, 1982; Nishida, 1968; Nishida, *et al.*, 1979; Norikoshi, 1983; Takahata, *et al.*, 1984; Uehara, *et al.*, 1992). This paper is intended to update this information and to reveal the characteristics of chimpanzees as predators of mammals while taking into account the similarities and differences among chimpanzee populations. Recently, primatologists have stressed the importance of "social" (Stanford, *et al.*, 1994a, b) or "demographic" (Mitani and Watts, 1999) influences on the hunting behavior of chimpanzees. Their ideas are based on the assumption that chimpanzees would predict success potential according to party size and composition when encountering prey monkeys and then decide whether or not to hunt them. Although one of us succeeded in obtaining comparable data from this viewpoint for the first time at Mahale, we do not go into too much detail here because the issue will be treated systematically elsewhere (Hosaka, in prep.).

2. METHODS

Chimpanzees of the M group in the Mahale Mountains, Tanzania were studied in the years 1991-1995 (Table 7). The study covered about two-thirds of their 30 km² home range in the Kasoje area along the eastern shore of Lake Tanganyika. The habitat is characterized as a mosaic of various vegetations including semi-deciduous forests and deciduous woodlands. Two discernible seasons are present in Mahale: the rainy season (from October to mid-May) and the dry season (from mid-May to September), but inter-annual shifts are common. Rainfall at Kansyana Camp, which is located near the center of the M group home range, was 1,835.5 mm on average in the period 1973-1988 (Takasaki, *et al.*, 1990). For more details, see Collins and McGrew (1988), Nishida (1972), Nishida (1990a), Nishida and Uehara (1981), and Uehara and Ihobe (1998).

The size of the M group shrank from 99 in 1981 to 62 in 1995. Although this phenomenon could not be explained clearly, some possibly relevant information is available, such as lion predation in 1989-1990 (Tsukahara, 1993) and an epidemic of an influenza-like disease in October 1993 (Hosaka, 1995a; Hosaka, et al., 2000). In contrast, the number of adult males was relatively stable in M group and varied from eight to ten.

Table 7. Study periods during which updated information was gathered

Study year	Study period	Observers[a] contributing information on predation	Study days	No. of group days[b]
1990	2 Feb – 31 Mar 1991	MB, MH, HT	33	17
1991	1 Apr 1991 – 14 Mar 1992	MB, MH, KH, MAH, RK, MhN, TN, RN	263	206
1992	7 Apr – 5 Nov 1992 26 – 31 Mar 1993	MB, KH, HH, MhN, TN, KN, RN, SU	194	156
1993	1 Apr 1993 – 31 Mar 1994	RH, KH, KK, RK, AM	268	221
1994	1 Apr 1994 – 31 Mar 1995	MB, RK, JM, McN, TN, LT	272	213
1995	1 Apr – 28 Nov 1995	NI, RK, MhN, TN	168	123

[a] MB: Moshi Bunengwa; MH: Miya Hamai; HH: Hitoshige Hayaki; RH: Rashidi Hawazi; KH: Kazuhiko Hosaka; MAH: Michael A. Huffman; NI: Noriko Itoh; KK: Kenji Kawanaka; RK: Rashidi Kitopeni; AM: Akiko Matsumoto-Oda; JM: John C. Mitani; McN: Michio Nakamura; MhN: Miho Nakamura; TN: Toshisada Nishida; KN: Kohshi Norikoshi; RN: Ramadhani Nyundo; HT: Hiroyuki Takasaki; LT: Linda A. Turner; SU: Shigeo Uehara.
[b] This term was defined by Uehara, et al. (1992) as the number of days on which more than ten chimpanzees were seen. Predation was recorded only on group days in this study.

Data were gathered by a total of 15 researchers and four Tanzanian assistants (Table 7). Two researchers, Hosaka and Nishida, who contributed 47.5% of recorded data on the number of eaten prey (n=29: Table 9), mainly studied social behavior and communication among adult males by the focal sampling method (Martin and Bateson, 1993). However, each time either hunting or meat-sharing occurred, the two changed to the *ad libitum* sampling method to confirm information on prey, predators, and meat-sharing.

For comparison with studies of predation of monkeys in other chimpanzee populations, some analyses were based only on Hosaka's data with the following definitions of terms. *Encounter* was counted when at least one chimpanzee reached the group beneath the trees in which prey was visible. Two encounters were discriminated if temporally more distant than 30 min. or spatially more distant than 100 m, which sufficiently exceeds the estimated group spread of prey monkey species at Mahale (Uehara and Ihobe, 1998). *Hunt* was counted when at least one chimpanzee either began to climb a tree to approach prey or rushed toward prey on the ground. *Success* in hunts were counted when at least one chimpanzee captured prey. In this study, the encounters with red colobus (n=117) were systematically recorded by Hosaka only during the study years 1991 and 1993.

This study was based only on direct observations of chimpanzee behavior. Thus, when citing data from previous reports for this study, fecal data were neglected. Sometimes chimpanzees were observed carrying skin and/or bone that had been consumed, but the hunt itself was not witnessed. We treat such cases by counting them only when no hunts had been recorded either on the same or the previous day. Although ten cases of cannibalism of infants have been documented (Hamai, *et al.*, 1992; Hosaka *et al.*, 2000; Kawanaka, 1981; Kitopeni, unpublished; Nishida, 1998; Nishida and Kawanaka, 1985; Norikoshi, 1982; Takahata, 1985), these were eliminated from the analyses presented here.

3.   RESULTS

3.1   Prey Profile and Predation Frequency

Red colobus (*Colobus badius*) was the largest proportion (83.3%, n=245 in 1990-1995, Table 8) of the mammals eaten by Mahale chimpanzees. This tendency remained constant during the entire study period. Blue duiker (*Cephalophus monticola*) was the second largest proportion (7.5%), although the number of eaten duikers peaked in 1991 and decreased every year to zero in the last two study years. Other prey species included red-tailed monkeys (*Cercopithecus ascanius*), vervet monkeys (*Cercopithecus aethiops*), and blue monkeys (*Cercopithecus mitis*). No predation on yellow baboon (*Papio cynocephalus*) was recorded here, as in the previous reports. Nakamura (1997) observed the first and only case of predation of an infant baboon in 1996.

Figure 23 illustrates the change in prey profile throughout the 30-year study of the M group. Most remarkable is the drastic increase in the proportion of prey that were red colobus. In the initial stage (1996-1981), red colobus was eaten only at the third highest proportion (14%, n=50 in 1966-1981), following bushbuck (*Tragelaphus scriptus*, 24%), and blue duiker (22%). Furthermore, the total number of prey eaten in each study period greatly increased. The sample size of the second period (1983-1989) was double that of the first stage. The sample size of the final stage (1990-1995) was triple the second one. To demonstrate how this change occurred from 1981 to 1995, predation frequencies of all mammals, red colobus, and blue duiker are estimated per year (Table 9). The increase in predation frequency of mammals was correlated with that of red colobus (Spearman r=0.89, p<0.005) but not with that of blue duiker (r=0.035, p=0.9).

Table 8. Mammals eaten by chimpanzees of Mahale M group (1990[a]-1995)

Prey species	1991	1992	1993	1994	1995	Total	
Primates							
Red colobus	45	47[b]	57	68	28	245	(83.3%)
Red-tailed monkey	2	1	1	0	0	4	(1.4%)
Vervet monkey	0	0	0	1	0	1	(0.3%)
Blue monkey	0	1	0	0	0	1	(0.3%)
Artiodactyla							
Blue duiker	12	7	3	0	0	22	(7.5%)
Bushbuck	0	7	1[b]	2	0	10	(3.4%)
Bushpig	3	0	1	0	1	5	(1.7%)
Rodentia							
Cane rat	0	0	1	0	0	1	(0.3%)
Squirrel (unidentified)	1	0	0	2	2	5	(1.7%)
Total	63	63	64	73	31	294	(100%)

[a] Study year 1990 had only one case of predation of a juvenile blue duiker on 22 Feb 1991.
[b] One case of scavenging was included in each cell.

Predation frequency of mammals (r=-0.76, p>0.05) and red colobus (r=0.84, p>0.01) increased despite the decrease in size of the M group, but that of blue duiker fluctuated irregularly (r=0.18, p=0.58). In contrast, the number of adult males, relatively stable despite the decrease in the M group's size, did not show any correlation to predation frequency (mammals: r=0.01, p=0.76; red colobus: r=0.08, p=0.80; blue duiker: r=0.08, p=0.81).

The number of researchers varied among the years, but this number was not correlated with the predation frequency of red colobus (r=0.25, p=0.43).

Considering that predation frequencies in Table 9 were biased by possible differences in observational efforts, another estimate of predation frequency of red colobus was made on the basis of Hosaka's observation time (1991: 894.2 hr; 1993: 1230.8 hr). Thus, yearly predation frequency of red colobus, estimated as the number of kills in observed hunts/total observation time x 12 x 365, is 153 according to the data combined from 1991 and 1993 (1991: 122; 1993: 174). Even if the data are restricted to focal sampling (1991: 647.9 hr; 1993: 942.0 hr), the result is very similar (1991 and 1993 combined: 146; 1991: 101; 1993: 177).

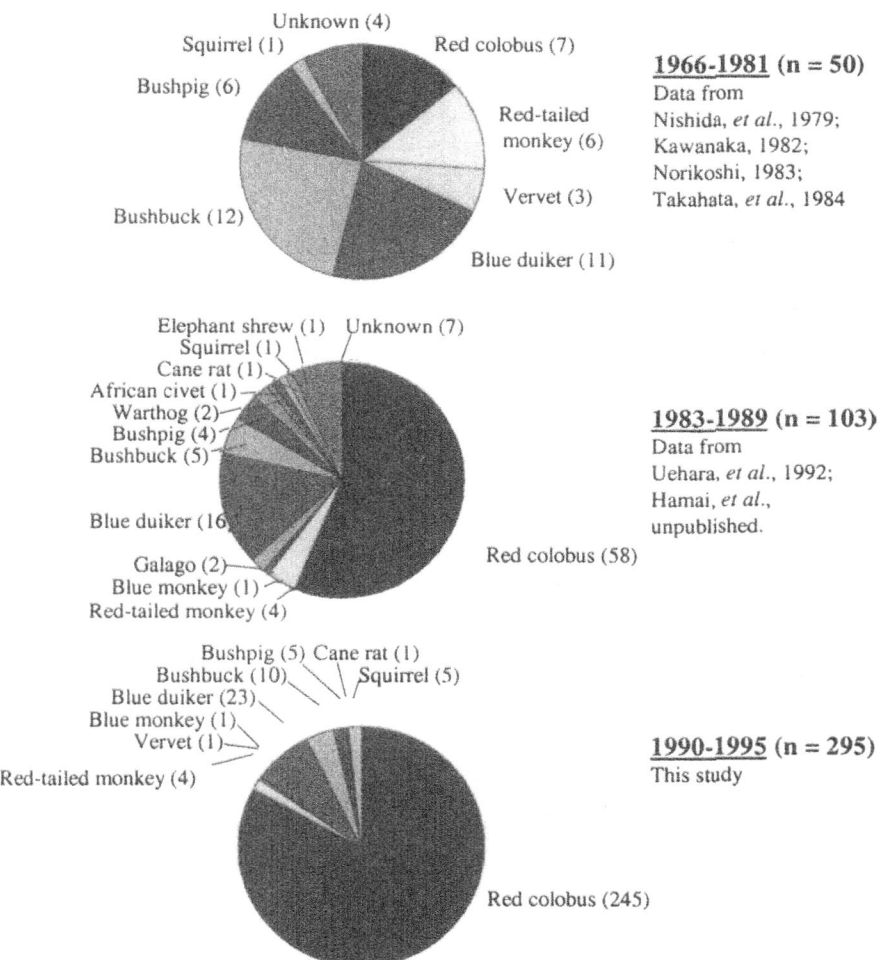

Figure 23. Change in prey profile throughout 30-year study of M group

Table 9. Size and composition of M group and predation frequency of mammals

Study year	Size of M group	No. of adult males	No. of adult males	Predation frequency[c] of		
				mammals	colobus	duiker
1981	99	9	39	71	13	18
1983	97	9	36	71	43	0
1985	99	10	39	32	23	0
1987	89	8	33	61	23	23
1988	96	10	35	56	20	13
1989	86	10	32	88	63	11
1991	79	10	26	112	82	21
1992	81	10	25	147	110	16
1993	76	10	23	106	94	5
1994	67	9	20	125	117	0
1995	62	8	19	92	83	0

[a] Defined as the number of individuals at the end of December each year.
[b] Determination of adulthood in chimpanzees followed Hiraiwa-Hasegawa, et al. (1984).
[c] (Number of prey/number of group days) x 365.

3.2   Comparison to a small-sized unit-group

At Mahale, other data are available from the K group, the now extinct Mahale unit-group that was studied until the beginning of the 1980s (Nishida, et al., 1979; Norikoshi, 1983; Takahata, et al., 1984). During the 16-year period of 1966-1981, K group chimpanzees were observed preying on 20 mammals (three red colobus, one red-tailed monkey, one galago, nine blue duikers, two bushbucks, one bushpig, one giant rat, and two rock hyraxes), which is less than half of the amount from the data of the M group taken in the same period (50).

Uehara, et al. (1992) indicated that the small size of the K group (26-34 total, 5-6 adult males, 8-10 adult females in 1967-1973: Nishida, 1979) might account for this difference. Since arboreal monkeys are known to be killed by group hunting (Uehara, 1997; Uehara, et al., 1992), it is reasonable to expect that the large-sized M group ate monkeys more frequently than other prey. However, comparison between these two unit-groups in the same period yields no significant difference between the numbers of primates and artiodactyls (χ^2=0.21, 1 df, p=0.6).

3.3   Seasonality

There was distinct seasonal variation in the number of prey eaten per month (Figure 24) according to the data (n=431) pooled from Takahata, et al. (1984), Uehara, et al. (1992), and this study. The average number of prey eaten per month was calculated by dividing the sum of the numbers of prey by the number of months in which observations were made. Figure 24 also shows that chimpanzee predation at Mahale peaked around August and September. Generally, the meat-eating season corresponded to the entire dry season and the first half of the rainy season at Mahale. In addition, this pattern may depend on the predation of red colobus (n=309), as there was strong correlation between average numbers of prey and red colobus eaten per month (Spearman r=0.9, p<0.005).

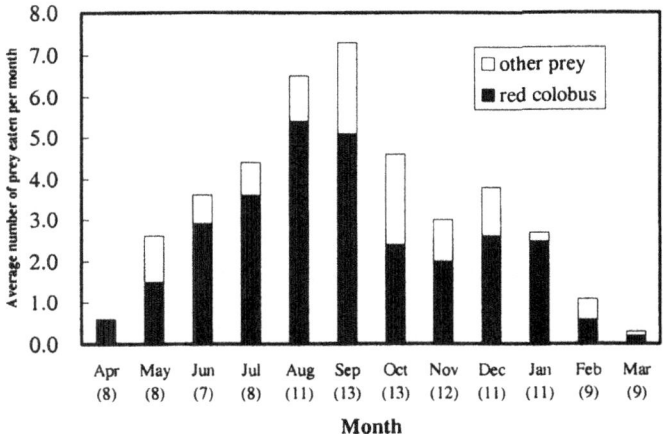

Figure 24. Seasonal variation in number of prey eaten per month (1979-1995). Figures in parentheses show numbers of months in which observations were made

3.4   Hunting Behavior

3.4.1   Encounter, hunt, success

In 1991 and 1993, a total of 117 encounters with red colobus were observed by Hosaka while following chimpanzees. In these encounters, hunts were undertaken 74 times (63.2%), and at least one monkey was captured in 44 cases (59.5%).

3.4.2   Differences in reactions to prey species

Chimpanzees showed different reactions according to the prey species they encountered. Small terrestrial mammals such as blue duiker and cane rat were seized abruptly upon being detected by chimpanzees (see also Takahata, *et al.*, 1984). Squirrels might be killed in a more exploratory way, as described in Huffman and Kalunde's (1993) report of a female chimpanzee that had captured a squirrel by probing with a branch in the tree hole where the prey hid itself.

Red colobus was killed mostly by groups of three or more chimpanzees. Interestingly, the same tendency was not seen when hunting red-tailed monkey, another abundant primate species at Mahale, which was captured only by one or two hunters (Table 10). Such a difference in chimpanzee reaction to prey species has not yet been scrutinized in other populations having sympatric arboreal monkeys (e.g., Boesch and Boesch, 1989; Mitani and Watts, 1999; Stanford, *et al.*, 1994a; but see Teleki, 1973).

Table 10. Difference in hunting group size according to prey monkey species (1991 and 1993)

Prey species	Hunting group size[a]					
	1		2		≥ 3	
	Hunt	Success	Hunt	Success	Hunt	Success
Red colobus	2[b]	0	3[c]	1	69[d]	43
Red-tailed	2[e]	0	3[f]	1	0	0

[a] Number of chimpanzees that encountered prey at the same time and undertook hunts (see also Boesch and Boesch, 1989).
[b] Hunts were undertaken out of five encounters. All hunters were adult males with no bystanders.
[c] Hunts were undertaken out of eight encounters. All hunters were adult males with no bystanders.
[d] Hunts were undertaken out of 104 encounters.
[e] In one case, the hunter was an infant female with nine bystanders (two adult males, one adolescent male, three adult females). In the other case, the hunter was an adult male without bystanders.
[f] In all cases, hunters were adult males. In two cases, there was an adult male bystander.

There were clear differences between small hunting groups (one or two hunters) and large hunting groups (three or more hunters). First, small hunting groups consisted mostly of adult males (see notes in Table 10), while large ones included individuals of all age/sex classes. In the latter, however, it was difficult to identify all hunters as they were dispersed to monitor the prey group as it spread out. Second, small hunting groups had few bystanders watching the hunters pursue the prey (see notes in Table 10), while in large hunting groups there were always bystanders monitoring the behaviors of hunters and prey from the ground. Finally, small hunting groups were quiet, never vocalizing or heard by other chimpanzees. In contrast, large hunting groups were noisy, emitting many vocalizations such as pant-hoots, pant-grunts, barks, and screams. Adult males often conducted charging displays, and bystanders barked at monkeys while the prey in turn emitted alarm calls and barked at predators below.

Although yellow baboons are abundant at Mahale (Uehara and Ihobe, 1998), we rarely observed encounters between them and chimpanzees. We frequently heard baboon vocalizations while we followed chimpanzees in the woodland area near the lake. Possibly they avoided each other, although the precise relationship between the two species is unknown. In one case in 1993, Hosaka observed an adult male chimpanzee rush into a dry riverbed with piloerection and conduct a charging display against baboons feeding on the fruit of a fig tree (*Ficus vallis-choudae*). The baboons were displaced, and the chimpanzee climbed the tree to feed.

3.4.3   Hunting methods

In this study, no systematic record on detailed hunting methods was carried out for large group hunting of red colobus. This is simply because we could not observe all of the hunters that chased prey in an area of 60 m in diameter at maximum (Uehara and Ihobe, 1998).

A typical pattern of group hunting of red colobus could be described as follows. Several hunters climb different trees to approach different prey, usually in a non-cooperative manner. When hunters reach a high point in the tree, some monkeys escape from the predators by jumping to the next tree, some continue to hide themselves with the foliage of leaves and vines, and some become frozen in the face of predators. Hunters attempt to search and capture such hidden or frozen monkeys while other bystanders beneath the trees sit or walk back and forth monitoring the behaviors of hunters and prey. If a monkey falls to the ground, one of these bystanders gets the opportunity to kill it.

In 1991-1995, we confirmed nine captures by such bystanders on the ground. In all cases, adult males captured the prey. In contrast, prey was captured in the tree in 45 other confirmed cases (captors: 24 adult males, 11 adolescent males, six adult females, three adolescent females, one juvenile female). Thus it was found that adult males adopt the "wait for a monkey to fall" tactic significantly more often than other age/sex individuals (Fisher's exact probability test, p>0.01).

Small hunting groups comprising one or two hunters showed different patterns (Table 11). Solitary hunts of red colobus (n=2) and red-tailed monkeys (n=2) were done by stalking the prey in trees; all cases resulted in failure. In one hunt of a red-tailed monkey, three adult males sat side by side watching a 30 m distant monkey. Two of them started to stalk the prey in single file and then they chased the prey in turns, resulting in failure. In three hunts (two of colobus, one of a red-tailed) in similar situations to the above, two adult males synchronized their actions toward a monkey, also resulting in failure. The remaining two hunts (one of a colobus, one of a red-tailed) were categorized as collaboration. Upon detecting prey, one adult male (subordinate) extended his hand to the other (alpha male) and embraced him by the shoulder with a grinning face (see ethogram: Nishida, *et al.*, 1999). This pattern was the same as support-seeking observed in conflicts among adult males (de Waal and van Hooff, 1981). Then the two males stalked the prey in single file. As soon as the subordinate male quickly climbed the tree, the target jumped down to the ground and the alpha male rushed to capture it.

All hunts by two hunters (n=6) took place in open woodland, whereas most hunts by three or more hunters were observed in forests. In addition, no monkey other than a single target was seen in each hunt.

Table 11. Hunting methods employed by small hunting groups (1991 and 1993)

Prey species	Size of hunting groups	Hunting methods[a]							
		Stalk		Chase in turns		Synchronization		Collaboration	
Red colobus	1	2	(0)	0	-	0	-	0	-
	2	0	-	0	-	2	(0)	1	(1)
Red-tailed	1	2	(0)	0	-	0	-	0	-
	2	0	-	1	(0)	1	(0)	1	(1)
Total		4	(0)	1	(0)	3	(0)	2	(2)

[a] See text for explanation. Figures in parentheses show numbers of successful hunts.

3.4.4   Multiple killing

Multiple killing (e.g. Boesch and Boesch, 1989; Mitani and Watts, 1999; Stanford, *et al.*, 1994a; Uehara, *et al.*, 1992) is defined here as killing two or more animals in a successful hunt. However, since encounters with red colobus were not recorded except by Hosaka, here we adopt a criterion of at least a one-hour interval to discriminate separate hunts.

Multiple killing was only observed in red colobus hunts. Exceptionally, three infant bushpigs, presumably of the same litter, were killed on 14 September 1991. In 1991-1995, 23.1% of all successful hunts resulted in multiple killing (Table 12). The average number of kills in a single hunt was 1.4 (range 1-9). Multiple killing accounted for 46.9% of all kills.

3.4.5   Age of prey

The chimpanzees of Mahale tended to prey on immatures (Table 13), although this cannot be confirmed until the data on age/sex proportions of each prey species is available. The proportion of immatures among red colobus eaten by the Mahale chimpanzees was 80.6% (n=155), which seemed similar to the data from Gombe (84%, n=241: Stanford, pers. comm. cited in Mitani and Watts, 1999). Taï chimpanzees did not appear to select immatures (47%, n=58: Boesch and Boesch, 1989). The Ngogo chimpanzees showed tendencies (66%, n=98: Mitani and Watts, 1999) intermediate to those of Mahale/Gombe and Taï.

Table 12. Multiple killing of red colobus

No. of kills per hunt	Study periods					
	Total	1991	1992	1993	1994	1995
One	130	27	27	31	29	16
Two	24	3	4	8	8	1
Three	8	1	1	2	3	1
Four	2	0	1	1	0	0
Five	2	0	1	0	1	0
Six	0	0	0	0	0	0
Seven	1	0	0	0	0	1
Eight	0	0	0	0	0	0
Nine	2	1	0	0	1	0
No. successful hunts	169	32	34	42	42	19
No. multiple killing hunts	39	5	7	11	13	3
% hunts that resulted in multiple killing	23.1	15.6	20.6	26.2	31.0	15.8
No. kills in successful hunts	245	45	47	57	68	28
No. kills in multiple killing	115	18	20	26	39	12
% kills obtained in multiple killing	46.9	40.0	42.6	45.6	57.4	42.9
Average no. kills in successful hunts	1.4	1.4	1.4	1.4	1.6	1.5

Table 13. Age of prey eaten by chimpanzees (1990–1995)

Age classes[a] of prey	Prey species[b]							
	RC	RT	BL	VV	BD	BB	BP	SQ
Adult	30[c]	1	1	0	3	1[d]	0	0
Adolescent	38[e]	0	0	0	1	0	0	0
Juvenile	48	1	0	0	5	2[f]	0	0
Infant	39	0	0	1	6	2	4	2
Unknown	89	2	0	0	8	5	1	3
Total	245	4	1	1	23	10	5	5

[a] Presumed.

[b] RC: red colobus; RT: red-tailed monkey; BL: blue monkey; VV: vervet monkey; BD: blue duiker; BB: bushbuck; BP: bushpig; SQ: squirrel.

[c] Sex: one male, seven females, 22 unknown. A confirmed case of scavenging an adult female is included.

[d] A presumed case of scavenging an adult female bushbuck.

[e] Sex: three females, 35 unknown.

[f] A confirmed case of scavenging a juvenile bushbuck is included.

3.4.6   Age/sex class of captors

In 48.6% of prey killed by the Mahale chimpanzees, captors were confirmed by direct observations (Table 14). Adult males captured mammals by the largest proportion (49.3%). In particular, adult males captured red colobus more abundantly (51.3%) than did adolescent males (20.5%) or adult females (22.2%). Although the possibility that adult males hunt monkeys while adult females hunt artiodactyls has been suggested (Takahata, *et al.*, 1984; Uehara, *et al.*, 1992), this study did not support it (χ^2=1.5, 1 df, p=0.22).

Table 14. Age/sex class of captors by prey species (1990–1995)

Age/sex class [a] of captors	Prey species[b]									
	RC	RT	BL	VV	BD	BB	BP	SQ	CR	Total
Adult M	60	2	0	1	4	2	0	0	1	70
Adol. M	24	0	0	0	0	0	0	1	0	25
Juve. M	1	0	0	0	1	0	0	0	0	2
Adult F	26	0	0	0	4	1	1	0	0	32
Adol. F	5	0	0	0	3	0	0	1	0	9
Juve. F	1	0	0	0	1	0	0	2	0	4
Unknown[c]	127	2	1	0	10	5	4	1	0	150
Total	244	4	1	1	23	8	5	5	1	292

[a] Adol.: adolescent; Juve.: juvenile. M: male; F: female.
[b] RC: red colobus; RT: red-tailed monkey; BL: blue monkey; VV: vervet monkey; BD: blue duiker; BB: bushbuck; BP: bushpig; SQ: squirrel; CR: cane rat.
[c] Excluding confirmed and presumed cases of scavenging.

3.4.7   Rank in number of captures

Boesch and Boesch (1989) and Stanford, *et al.* (1994a) suggested that there are persistent hunters or good hunters and that their presence in the group may influence hunting decisions or hunting success. In this study, among 11 individuals whose numbers of captures of red colobus ranked 1 to 10, there were seven adult males, two adolescent males, one adult female, and one adolescent female (Table 15).

Several features are noteworthy in Table 15. First, adult males that experienced alpha status during this study tended to rank high in number of captures (Kalunde, Ntologi, and Nsaba: see notes in Table 15). Second, two adolescent males (Alofu and Dogura) ranked high, but also often lost carcasses due to theft (no. stolen carcass/no. captures: Alofu 6/8, Dogura 5/7; also see below). Finally, there were two females (Fatuma, Linda) who ranked within ten.

Table 15. Rank of chimpanzees based on recorded captures of red colobus (1991–1995)

Rank	Name	Age/sex class	No. captures[a]
1	Kalunde[b]	adult male	15
2	Toshibo	adult male[c]	13
3	Ntologi[d]	adult male	9
4	Alofu	adolescent male	8
5	Dogura	adolescent male	7
6	Fatuma	adult female	6
	Nsaba[e]	adult male	6
8	Jilba	adult male	4
	Linda	adolescent female[f]	4
10	Aji	adult male	3
	Bakali[g]	adult male	3

[a] Only individuals with three or more recorded captures were listed here. Twenty-three more individuals (ten males, 13 females) were confirmed as captors at least once.
[b] Six captures done as alpha male during 10-month period from 16 Mar 1991 to 19 Jan 1992.
[c] Still categorized as an adolescent male in 1991.
[d] All captures done as alpha male during three-year and three-month period from 20 Jan 1992 to 19 Apr 1995.
[e] Three captures done as alpha male during seven-month period from 20 Apr 1995 to 28 Nov 1995.
[f] Grew up to be categorized as an adult female in 1995.
[g] Observed only in 1991 and 1992, as he presumably died in March 1993.

Table 16. Stealing of carcasses from captors (1991-1995)

	Age[a]/sex class of captors					
	Male			Female		
	Adult (12)[b]	Adol. (10)	Juve. (6)	Adult (30)	Adol. (13)	Juve. (9)
No. captures	70	25	2	32	9	4
Average no. captures per individual	5.8	2.5	0.3	1.1	0.7	0.4
No. carcasses stolen	10	15	1	13	3	2
Age/sex class of thieves						
Adult male[c]	7	11	1	12	3	2
Adolescent male	0	2	0	1	0	0
Juvenile male	0	0	0	0	0	0
Adult female	3	2	0	0	0	0
Adolescent female	0	0	0	0	0	0
Juvenile female	0	0	0	0	0	0

[a] Adol.: adolescent; Juve.: juvenile.
[b] Figures in parentheses show the numbers of individuals who belonged to each age/sex class for more than six months during the study periods.
[c] Alpha males comprised more than half (19 out of 36) of thefts by adult males.

3.4.8   Theft

Carcasses were often stolen from other individuals, usually within two or three minutes after hunters captured prey.

Table 16 shows the number of carcasses that were stolen from hunters of each age/sex class by thieves of each age/sex class. The average adult male most-frequent captor (5.8 captures/individual) lost 14.3% of his carcasses to other adult males (7/10) or adult females (3/10). The average adolescent male second-most-frequent captor (2.5 captures/individual) lost 60.0% of his carcasses, mostly to adult males (11/15). The average adult female third-most-frequent captor (1.1 captures/individual) lost 40.6% of her carcasses, mostly to adult males (12/13).

4.   DISCUSSION

4.1   Chronological Change in Predation Frequency of Red Colobus

This study revealed the recent trends in predation of mammals by the Mahale chimpanzees compared with information previously reported at the same site (Kawanaka, 1982; Nishida, *et al.*, 1979; Takahata, *et al.*, 1984; Uehara, *et al.*, 1992). First, the proportion of red colobus among all eaten prey (83.3%) drastically increased. Therefore, we can conclude that the current Mahale chimpanzees have the same level of prey selectivity as their counterparts in other populations (Gombe: 82%, Stanford, *et al.*, 1994a; Taï 78%, Boesch and Boesch, 1989; Ngogo, 91%, Mitani and Watts, 1999). Second, the frequency of predation of mammals also dramatically increased. As correlation analyses in this study showed, both changes were simply produced by the increased frequency of predation on red colobus.

Why did the Mahale chimpanzees begin to hunt red colobus more frequently than before? One answer might be relevant to the history of Mahale (Nishida, 1990a). In 1974, following a government edict, most villagers moved out of Kasoje and abandoned farming, burning, and hunting. In 1985, Mahale became the eleventh national park in Tanzania. Uehara and Ihobe (1998) suggested that red colobus and red-tailed monkeys gradually increased their populations in Mahale. It is likely that the increase in faunal availability may have benefited chimpanzees as predators of mammals. An alternative explanation is based on floral change. Although no systematic census of flora was carried out, it seems probable that forest recovery occurred, increasing fruit availability. Chimpanzees are known to increase party size in response to the seasonal abundance of fruits (Chapman, *et al.*, 1995; Doran, 1997). In addition, as Stanford, *et al.*, (1994a, b), Mitani and Watts (1999), and Hosaka (1995b; in prep) have demonstrated, chimpanzees hunt red colobus more frequently as party size increases. Therefore, the possible recovery of fruit plants might have led to an increase in daily party size and to higher rates of predation on red colobus.

Besides such ecological explanations, methodological changes might have influenced the differences between the previous data and this study's data. Around 1980, focal (or focal subgroup) sampling methods became common among researchers, and thereafter chimpanzees have been followed into the forests routinely (see papers in Nishida, 1990b). This change in methodology possibly increased the frequency of researchers observing chimpanzees encounter and hunt red colobus. Wrangham and Bergmann Riss (1990) gave a similar explanation for the sudden increase in predation frequency around 1973 in the research at Gombe. Although this may explain part of the change in Mahale from the initial period (1966-1981) to the second (1983-1989), it cannot account for the change from the second to the final period (1990-1995). The number of researchers was also found to be unrelated to predation frequency of red colobus, and so this cannot account for the chronological change observed.

Stanford, *et al.* (1994a) suggested that the Gombe chimpanzees increased the hunt rate of red colobus during the 1980s due to the increase in the number of adult males in the unit-group. However, the number of adult males in M group is not a viable explanation for the Mahale data (Table 9).

4.2   Prey Selectivity

The question remains as to why the Mahale chimpanzees select red colobus as prey rather than red-tailed monkey, the second most abundant monkey inhabiting forests (Uehara and Ihobe, 1998). The answer to this question might lead to a hypothesis explaining the high prey selectivity for red colobus in other populations.

Some clues may be obtained by comparing the two prey species in reference to group structure and anti-predator behavior. The red colobus' large group size and wide group spread may lead to high detectability for predators (Teleki, 1973), although living in large groups is often considered to benefit prey by improving detection of predators in many species (Bertram, 1978; van Schaik, 1983). That is, a red colobus group may be too large and spread too widely to prevent chimpanzee predation, although some moderate size would reduce detection by predators in that sparse groups would be detected less frequently than isolated individuals.

This hypothesis was not strongly supported by the present information from Mahale. According to the results of Uehara and Ihobe's (1998) route census in forests, there was only a small difference in the frequency of encounters with colobus versus red-tailed monkeys, despite other estimates that the group size of colobus (30) was larger than that of red-tailed (12) and that the group spread was wider for colobus (30 m in radius) than for red-tailed (10 m in radius).

The anti-predator behavior of the red colobus may be posed as an alternative to explain prey selectivity. Ihobe (1999) suggested that the Mahale red colobus adopts the "stand and defend" strategy against chimpanzee predators, which might be similar to that of its Gombe counterpart (Stanford, 1995). However, as several researchers at other study sites indicated (Busse, 1978; Mitani and Watts, 1999; Stanford, 1995, 1998a, b), mobbing by red colobus may not function well against chimpanzees hunting in groups (see below). Rather, red colobus may be favorable prey for chimpanzees in that it does not flee upon detecting predators. In contrast, red-tailed monkeys may flee quickly upon hearing chimpanzees. Hosaka (pers. obs.) noted that clumped groups of red-tailed monkeys were rarely seen nearby while following chimpanzees.

A different approach to prey selectivity is possible by considering "search image" (Endler, 1991). Taking account of the conservative nature of chimpanzees in selectivity of plant foods (Nishida and Uehara, 1983), it is possible that chimpanzees come to persist in searching the same prey species after they repeatedly succeed in hunting it and acquire a sufficient image about its distribution and behavior (see also Boesch and Boesch, 1989). "Hunting patrols" at Ngogo (Mitani and Watts, 1999) implies that chimpanzees detect prey with the aid of such a search image. No evidence is available for prey searching behavior at Gombe (Stanford, *et al.*, 1994a), but it is possible that the Gombe chimpanzees adjust their ranging pattern, with the expectation of meeting red colobus (Stanford, 1998a). At Mahale, there is also no direct evidence for a search image in chimpanzee behavior. However, Hosaka (in prep) found indirect evidence that chimpanzees encountered red colobus nonrandomly, as the encounters themselves showed temporal fluctuation corresponding to that of hunts.

4.3   Hunting Strategy

Although further study of prey ecology and behavior is needed, the reactions of chimpanzees to prey documented in this study suggest that chimpanzees may vary their hunting tactics according to the behaviors of prey and other participants in hunts.

When hunting in groups, adult males may adopt the "actively hunt," "wait for a monkey to fall," or "steal" tactics (Tables 14, 16). Adolescent males (also younger adult males) may adopt the "actively hunt and run away upon capture" tactic, because they often lose carcasses to dominant males and even to adult females (Table 16). Most adult females may adopt the "wait for someone to capture prey" tactic, expecting that they can share meat with carcass holders, while some of them exploit the "actively hunt" tactic (such as the number 6 hunter, Fatuma, Table 15).

A clique of adult males walking in proximity (Table 11) can also adopt the "coalition" tactic to pursue the hunt of a lone, elusive monkey by utilizing social skills developed to manipulate relationships among males. That is, the patterns of adult male coalition in hunting behaviors and in agonistic confrontations (de Waal, 1982, 1984) may have the same behavioral origin, as the observed patterns were quite similar.

Undoubtedly, the above hunting tactics oversimplify actual events by averaging them to identify age/sex class tendencies. It seems that who adopts which tactic depends on each individual's relationships with the other individuals in the group. For example, an adult male who is an alpha male's ally may adopt the "wait for someone to capture prey" tactic (following Nishida, *et al.*'s (1992) hypothesis that the alpha male preferentially shares meat with his allies). Depicting a scheme of hunting strategy, however, may stimulate further study on what individual participants do in hunts.

4.4   Group Hunting

It seems that group hunting by the Mahale chimpanzees reflects some important traits of chimpanzee-type social hunters. In other words, chimpanzees' group hunting may share the same social and psychological mechanism with their coalition behaviors among conspecifics.

Boesch (1994b, c) stressed the Gombe chimpanzees' "fear" when confronted with red colobus aggression such as mobbing and male defense (see also Busse, 1977; Stanford, 1995, 1998a, b). He insisted that it is easy for a Gombe hunter to avoid such danger and select immature prey even when hunting uncoordinatedly with other individuals. In contrast, such hunting would have a higher energetic cost for a Taï hunter because of the high canopy and dense vegetation there. Taï chimpanzees cooperate in hunts so as to achieve high success, consequently overcoming their fear of the colobus.

Our updated data indicate that group hunting at Mahale is not characterized by cooperation (Boesch and Boesch, 1989). It seems correct that the Mahale chimpanzees avoid aggression from red colobus in that most captures were made on immatures, and only one adult male colobus was confirmed prey (Table 13).

However, our data suggest that group hunting, in itself, helped chimpanzees to overcome "fear" of red colobus. For instance, adult females and adolescent males could pursue and kill colobus only in group hunts (Table 14), while adult males did not need to hunt in groups to kill a red colobus (Tables 10, 11). Furthermore, group excitement observed only in group hunts may mitigate psychological difficulty inhibiting attack on red colobus.

What seems important in this respect is that bystanders as well as active hunters may play particular roles in group hunts. The term "bystander" refers here to any individual present at a hunt but not actively pursuing prey, similar to the definition used by Boesch and Boesch (1989). There is no reason to assume that an active hunter should always be the initiator of group hunts. Since Boesch and Boesch (1989) stated that the Taï chimpanzees initiated group hunts by "drumming and pant-hooting" and Mitani and Watts (1999) noticed a "hunting call" at Ngogo, it is plausible that some auditory cues, such as barking at prey, may function to induce others to participate in hunts. There is no evidence from any field site that such calls are emitted by active hunters. At Mahale, bystanders, including adult females and immatures, observed the behaviors of prey and active hunters and sometimes barked loudly (Hosaka, pers. obs.). Further study should reveal the roles played by bystanders in group hunts.

Then a new question arises: how does group hunting benefit a bystander? First, as Boesch and Boesch (1989) pointed out, a bystander can easily change to a hunter if a red colobus falls to the ground. In our study this occurred nine times, suggesting that it is a low-cost tactic exploited by adult males. Second, an adult male bystander may expect to steal a carcass once another chimpanzee kills a colobus in a tree and climbs down to the ground (Table 15). Finally and most importantly, bystanders can expect to share meat with carcass holders (Boesch, 1994a; Mitani and Watts, 1999; Nishida, et al., 1992; Teleki, 1973). Pursuing these possibilities may lead to a further study on some specific mechanism underlying the demographic effects on hunting behaviors suggested by Stanford, et al. (1994b) and Mitani and Watts (1999).

ACKNOWLEDGMENTS

This study was funded by grants under the Monbusho International Scientific Research Program (#03041046 and #07041138 to TN) and by the Japan Society for the Promotion of Science (Fellowship for Doctoral

Research to K.H. and A.M.). The Tanzania Commission for Science and Technology and the Serengeti Wildlife Research Institute permitted us to work in Mahale Mountains National Park. Edeus Massawe of the Mahale Mountains Wildlife Research Centre and his staff supported the fieldwork. We are extremely grateful to Moshi Bunengwa, Hitoshige Hayaki, Michael A. Huffman, Noriko Itoh, Kenji Kawanaka, Rashidi Kitopeni, John C. Mitani, Michio Nakamura, Miho Nakamura, Kohshi Norikoshi, Ramadhani Nyundo, Hiroyuki Takasaki, and Linda A. Turner for the use of their unpublished data. We also thank Yukio Takahata, Hiroshi Ihobe, Takayoshi Kano, and Juichi Yamagiwa for constructive comments and encouragement.

REFERENCES

Bertram, B.C.R., 1978, Living in groups: predators and prey. Pp. 64-96 in: (Eds. J.R. Krebs, and N.B. Davies), *Behavioural Ecology: An Evolutionary Approach*, Sunderland, MA: Sinauer.

Boesch, C., 1994a, Cooperative hunting in wild chimpanzees, *Anim. Behav.* 48: 653-667.

Boesch, C., 1994b, Hunting strategies of Gombe and Taï chimpanzees. Pp. 77-91 in: (Eds. R.W. Wrangham, W.C. McGrew, F.M.B. de Waal, and P.G. Heltne), *Chimpanzee Cultures*, Cambridge, MA: Harvard University Press.

Boesch, C., 1994c, Chimpanzees-red colobus monkeys: A predator-prey system, *Anim. Behav.* 47: 1135-1148.

Boesch, C., and Boesch, H., 1989, Hunting behavior of wild chimpanzees in the Taï National Park, *Am. J. Phys. Anthropol.* 78: 547-573.

Busse, C.D., 1977, Chimpanzee predation as a possible factor in the evolution of red colobus monkey social organization, *Evolution* 31: 907-911.

Busse, C.D., 1978, Do chimpanzees hunt cooperatively?, *Am. Nat.* 112: 767-770.

Butynski, T.M., 1982, Vertebrate predation by primates: A review of hunting patterns and prey, *J. Hum. Evolution* 11: 421-430.

Chapman, C.A., Wrangham, R.W., and Chapman, L.J., 1995, Ecological constraints on group size: An analysis of spider monkey and chimpanzee subgroups, *Behav. Ecol. Sociobiol.* 36: 59-70.

Collins, D.A. and McGrew, W.C., 1988, Habitats of three groups of chimpanzees (*Pan troglodytes*) in western Tanzania compared, *J. Hum. Evolution* 17: 553-574.

Doran, D., 1997, Influence of seasonality on activity patterns, feeding behavior, ranging, and grouping patterns in Taï chimpanzees, *Int. J. Primatol.* 18: 183-206.

Endler, J.A., 1991, Interactions between predators and prey. Pp. 169-196 in: (Eds. J.R. Krebs and N.B. Davies), *Behavioural Ecology: An Evolutionary Approach, 3rd ed.*, Oxford: Blackwell Scientific Publications.

Goodall, J., 1963, Feeding behaviour of wild chimpanzees: A preliminary report. *Symp. Zool. Soc. Lond.* 10: 39-48.

Goodall, J., 1986, *The Chimpanzees of Gombe: Patterns of Behavior*. Belknap: Harvard University Press.

Hamai, H., Nishida, T., Takasaki, H., and Turner, L.A., 1992, New records of within-group infanticide and cannibalism in wild chimpanzees, *Primates* 33: 151-162.

Hausfater, G., 1976, Predatory behavior of yellow baboons, *Behaviour* 56: 44-68.

Hiraiwa-Hasegawa, M., Hasegawa, T., and Nishida, T., 1984, Demographic study of a large-sized unit-group of chimpanzees in the Mahale Mountains, Tanzania: A preliminary report, *Primates* 25: 401-413.

Hosaka, K., 1995a, A single flu epidemic killed at least 11 chimps, *Pan Afr. News* 2(2): 3-4.

Hosaka, K., 1995b, Chimpanzee predation on red colobus in Mahale Mountains National Park, Tanzania, *Anthropol. Sci.* 103: 106.

Hosaka, K., Matsumoto-Oda, A., Huffman, M.A., and Kawanaka, K., 2000, Reactions to dead bodies of conspecifics by wild chimpanzees in the Mahale Mountains, Tanzania, *Primate Res* 16: 1-15 (in Japanese).

Huffman, M.A. and Kalunde, M.S., 1993, Tool-assisted predation on a squirrel by a female chimpanzee in the Mahale Mountains, Tanzania, *Primates* 34: 93-98.

Ihobe, H., 1992, Observations on the meat-eating behavior of wild bonobos (*Pan paniscus*) at Wamba, Republic of Zaire, *Primates* 33: 247-250.

Ihobe, H., 1999, Anti-chimpanzee-hunting strategy of red colobus monkeys at Mahale, Tanzania, *Anthropol. Sci.* 107: 71.

Kawanaka, K., 1981, Infanticide and cannibalism in chimpanzees, with special reference to the newly observed case in the Mahale Mountains, *Afr. Study Monogr.* 1: 69-99.

Kawanaka, K., 1982, Further studies on predation by chimpanzees of the Mahale Mountains, *Primates* 23: 364-384.

Kudo, H. and Mitani, M., 1985, New record of predatory behavior by the mandrill in Cameroon, *Primates* 26: 161-167.

Martin, P. and Bateson, P., 1993, *Measuring Behaviour, 2nd ed.*, Cambridge: Cambridge University Press.

Mitani, J.C., and Watts, D.P., 1999, Demographic influences on the hunting behavior of chimpanzees, *Am. J. Phys. Anthropol.* 109: 439-454.

Nakamura, M., 1997, First observed case of chimpanzee predation on yellow baboons (*Papio cynocephalus*) at the Mahale Mountains National Park, *Pan Afr. News* 4: 9-11.

Nishida, T., 1968, The social group of wild chimpanzees in the Mahali Mountains, *Primates* 9: 167-224.

Nishida, T., 1972, A note on the ecology of the red-colobus monkeys (*Colobus badius tephrosceles*) living in the Mahali Mountains, *Primates* 13: 57-64.

Nishida, T., 1979, The social structure of chimpanzees of the Mahale Mountains. Pp. 72-121 in: (Eds. D.A. Hamburg and E.R. McCown), *The Great Apes* Menlo Park, CA: Benjamin/Cummings.

Nishida, T., 1981, *The World of Wild Chimpanzees*, Chuokoronsha, Tokyo (in Japanese).

Nishida, T., 1990a, A quarter century of research in the Mahale Mountains. Pp. 3-35 in: (Ed. T. Nishida), *The Chimpanzees of the Mahale Mountains: Sexual and Life History Strategies*, Tokyo: University of Tokyo Press.

Nishida, T. (Ed.), 1990b, *The Chimpanzees of the Mahale Mountains: Sexual and Life History Strategies*, Tokyo: University of Tokyo Press.

Nishida, T., 1998, Deceptive tactic by an adult male chimpanzee to snatch a dead infant from its mother, *Pan Afr. News* 5: 13-14.

Nishida, T., and Kawanaka, K., 1985, Within-group cannibalism by adult male chimpanzees, *Primates*, 26: 274-284.

Nishida, T., and Uehara, S., 1981, Kitongwe names of plants: A preliminary listing, *Afr. Study Monogr.* 1: 109-131.

Nishida, T., and Uehara, S., 1983, Natural diet of chimpanzees (*Pan troglodytes schweinfurthii*): Long-term record from the Mahale Mountains, Tanzania, *Afr. Study Monogr.* 3: 109-130.

Nishida, T., Uehara, S., and Nyundo, R., 1979, Predatory behavior among wild chimpanzees of the Mahale Mountains, *Primates* 20: 1-20.

Nishida, T., Hasegawa, T., Hayaki, H., Takahata, Y., and Uehara, S., 1992, Meat-sharing as a coalition strategy by an alpha male chimpanzee? Pp. 159-174 in: (Eds. T. Nishida, W.C. McGrew, P. Marler, M. Pickford, and F.B.M. de Waal), *Topics in Primatology, vol. 1: Human Origins*, Tokyo: University of Tokyo Press.

Nishida, T., Kano, T., Goodall, J., McGrew, W.C., and Nakamura, M., 1999, Ethogram and ethnography of Mahale chimpanzees, *Anthropol. Sci.* 107: 141-188.

Norikoshi, K., 1982, One observed case of cannibalism among wild chimpanzees of the Mahale Mountains, *Primates* 23: 66-74.

Norikoshi, K., 1983, Prevalent phenomenon of predation observed among wild chimpanzees of the Mahale Mountains, *J. of the Anthro. Society of Nippon* 91: 475-479.

Rose, L.M., 1997, Vertebrate predation and food-sharing in *Cebus* and *Pan*, *Int. J. Primatol.* 18: 727-765.

van Schaik, C.P., 1983, Why are diurnal primates living in groups?, *Behaviour* 87: 120-144.

Stanford, C.B., 1995, The influence of chimpanzee predation on group size and anti-predator behavior in red colobus monkeys, *Anim. Behav.* 49: 577-587.

Stanford, C.B., 1998a, *Chimpanzee and Red Colobus: The Ecology of Predator and Prey.* Cambridge, MA: Harvard University Press.

Stanford, C.B., 1998b, Predation and male bonds in primate societies, *Behaviour* 135: 513-533.

Stanford, C.B., Wallis, J., Matama, H., and Goodall, J., 1994a, Patterns of predation by chimpanzees on red colobus monkeys in Gombe National Park, 1982-1991, *Am. J. Phys. Anthropol.* 94: 213-228.

Stanford, C.B., Wallis, J., Mpongo, E., and Goodall, J., 1994b, Hunting decisions in wild chimpanzees, *Behaviour* 131: 1-18.

Strum, S.C., 1981, Processes and products of change: Baboon predatory behavior at Gilgil, Kenya. Pp. 255-302 in: (Eds. R.S.O. Harding and G. Teleki), *Omnivorous Primates* New York: Columbia University Press.

Takahata, Y., 1985, Adult male chimpanzees kill and eat a male newborn infant: Newly observed intragroup infanticide and cannibalism in Mahale National Park, Tanzania, *Folia Primatol.* 44: 161-170.

Takahata, Y., Hasegawa, T., and Nishida, T., 1984, Chimpanzee predation in the Mahale Mountains from August 1979 to May 1982, *Int. J. Primatol.* 5: 213-233.

Takasaki, H., Nishida, T., Uehara, S., Norikoshi, K., Kawanaka, K., Takahata, Y., Hiraiwa-Hasegawa, M., Hasegawa, T., Hayaki, H., Masui K., and Huffman, M.A., 1990, Summary of meteorological data at Mahale research camps, 1973-1988. Pp. 291-300 in: (Ed. T. Nishida), *The Chimpanzees of the Mahale Mountains: Sexual and Life History Strategies*, Tokyo: University of Tokyo Press.

Teleki, G., 1973, *The Predatory Behavior of Wild Chimpanzees*, Lewisburg: Bucknell University Press.

Tsukahara, T., 1993, Lions eat chimpanzees: The first evidence of predation by lions on wild chimpanzees, *Am. J. Primatol.* 29: 1-11.

Uehara, S., 1997, Predation on mammals by the chimpanzee (*Pan troglodytes*), *Primates*, 38: 193-214.

Uehara, S. and Ihobe, H., 1998, Distribution and abundance of diurnal mammals, especially monkeys, at Kasoje, Mahale Mountains, Tanzania, *Anthropol. Sci.* 106: 349-369.

Uehara, S., Nishida, T., Hamai, M., Hasegawa, T., Hayaki, H., Huffman, M.A., Kawanaka, K., Kobayashi, S., Mitani, J.C., Takahata, Y., Takasaki, H., and Tsukahara, T., 1992, Characteristics of predation by the chimpanzees in the Mahale Mountains National Park, Tanzania. Pp. 143-158 in: (Eds. T. Nishida, W.C. McGrew, P. Marler, M. Pickford, and F.B.M. de Waal), *Topics in Primatology, Vol. 1: Human Origins*, Tokyo: University of Tokyo Press.

Utami, S.S. and van Hooff, J.A.R.A.M., 1997, Meat-eating by adult female Sumatran orangutans (*Pongo pygmaeus abelii*), *Am. J. Primatol.* 43: 159-165.

de Waal, F.B.M., 1982, *Chimpanzee Politics: Power and Sex Among Apes*, New York: Harper and Row.

de Waal, F.B.M., 1984, Sex differences in the formation of coalitions among chimpanzees, *Ethol. Sociobiol.* 5: 239-255.

de Waal, F.B.M., and van Hooff, J.A.R.A.M., 1981, Side-directed communication and agonistic interactions in chimpanzees, *Behavior* 77: 164-198.

Wrangham, R.W. and Bergmann Riss, E., 1990, Rates of predation on mammals by Gombe chimpanzees, 1972-1975, *Primates* 31: 157-170.

Chapter 9

REPRESENTATIONAL CAPACITIES IN CHIMPANZEES: NUMERICAL AND SPATIAL REASONING

S.T. Boysen and V.A. Kuhlmeier
Comparative Cognition Project, Dept. of Psychology, The Ohio State University, Columbus, OH 43210-1222

1.   INTRODUCTION

What counts to a chimpanzee? Since 1984, we have explored the capacity for counting and other complex cognitive skills in chimpanzees, a species with whom humans clearly share a special evolutionary heritage. However, demonstrations of a variety of cognitive capacities in chimpanzees, including numerical abilities, seem to raise intellectual hackles reminiscent of the animal language debate a decade ago (Boysen and Capaldi, 1993; also see Davis and Perusse, 1988 for a review of the animal counting literature). Barrow (1992) proposed a series of questions that address whether the human propensity for mathematical concepts might be innate:

- Does the human brain contain such a natural structure "hard-wired" into its make-up in some way?
- Does the human mind therefore possess a natural intuition for simple mathematical concepts?
- If yes, does this mean that our mathematical picture of the physical world is primarily a mind-imposed projection of our own mental structure upon the outside world?
- Did we discover counting or invent it?
- Could we have missed inventing counting and developed it in a literate but enumerate fashion?

131

- Did an intuition for counting arise all over the world wherever there was language and society of any sort, or was it come upon rather infrequently and then only by those in the midst of sophisticated cultural developments?
- Finally, what of counting itself? It appears to be necessary for mathematics, but is it sufficient to guarantee the sophisticated abstract notion of number and structure that forms the basis of the modern scientific picture of the world?

In examining the historical evidence about the emergence of counting in human cultures, Barrow (1992) concedes that, even if certain events or facets of the human mind were responsible for the development of mathematics, all we know is that they were sufficient, not necessary, to permit its development.   There is little question some members of the Academy continue to claim some type of intrinsic relationship between language and counting. We know, however, that all human cultures did not, and some still do not, have counting systems, although all possess spoken languages. Spoken language probably predated the origin of counting, and thus any natural propensity the brain might possess for certain patterns of thought likely evolved with bias towards effective linguistic or gestural communication, rather than counting. However, the evolution of a numerical sense would likely be a by-product of selection for some other primary attribute, perhaps language, or another neutral evolutionary development. Further, if the world were intrinsically mathematical, we should expect to find the signature of that feature "imprinted upon our minds" because of the selective advantage that is bestowed by a true perception of reality (Barrow, 1992).

We agree that having a notion of quantity is a long way from the types of abstract reasoning that are required for mathematics, most of which are very recent additions to human culture. However, a range of investigations in ethology and animal behavior indicates that a variety of nonhuman species deal with the basic abstractions of time, space, and number. For example, Gallistel (1989) noted that many animals make routine use of spatial cues and cognitive maps to find their way about during foraging, as was documented in the classic studies of the digger wasp. In studies of timing, rats (among other species) have been shown to represent when during the day particular events occur, represent temporal intervals, and keep a running count or estimation of the number of events that can then be compared to a standard stored in memory (e.g., Meck and Church, 1984). Numerous studies of pigeons, fish, and ducks have also explored nonhuman representation of rate (number divided by time), including relative and absolute rates. The assumption that other species can represent both number and temporal intervals, and perform operations that are formally analogous to division with these representations, could explain the fine tuning of some species' behavior, for example with respect to observed rates of prey occurrence. Similarly, animal navigation studies with birds and honeybees indicate that the calculations necessary for these kinds of spatial skills are fundamental behavioral processes for many animals who must depart from, and return to, a fixed point in their environment. A modern behavioral scientist might understand "mind" to be brain circuits that are tuned to locations of objects in egocentric space, or an animal's orientation and position in geometric space, and accompanying endogenous rhythmic processes that indicate time. The wealth of archival experimental data are consistent with the view that natural selection has shaped brains so that there exists a kind of brain/world parallelism of formal structure.

Among nonhuman primates, the chimpanzee represents a significant and appropriate species for the study of cognitive capacity, including the evaluation of numerical capabilities. Though certainly worthy of study in their own right, characterizations of chimpanzee skills elicit the inevitable comparisons with human capabilities. The past three decades of research with apes and symbol use, while generating controversy and forcing a reappraisal of our definition of humanness and language, support the conclusion that under some tutorial circumstances, apes can acquire facility with symbol systems that share some features of language-like processing and representation. It was with this background in mind that we initially conducted studies of numerical skills and counting with young chimpanzees. We have since expanded the range of questions addressed with respect to the nature of representations and cognitive capacities in chimpanzees. Details of training procedures for our counting work have been presented elsewhere (see Boysen, 1993; Boysen and Berntson, 1989).

The one number-based task that our animals failed to perform optimally has provoked more interest than some of their prior successes. The task was fairly straightforward, from our perspective, yet revealed something about the chimpanzee's preparedness to organize and respond to quantity that we would not have predicted from prior performance or experiences with them as sophisticated information-processors. Though originally planned as a simple task through which we hoped to study deception, the chimpanzees' inability to acquire an understanding of the task parameters encouraged us to take a closer look. The rules were simple. Two chimpanzees cooperated in the task, with one chimpanzee given the responsibility as the Selector while her partner was a passive Recipient. Two bowls containing varying amounts of candy were presented in comparison arrays of one versus two candies, one versus four, and one versus six. The chimpanzee's Selector role was to choose a dish, after which the teacher/experimenter intervened and gave the contents of the selected dish to the other chimpanzee. The candy in the second dish was then given to the Selector. If the chimpanzees came to recognize the rules of food division in this task, it was to their advantage to select the dish that had the smaller number of candies so that they might reap the larger, remaining amount. Many trials and sessions later, including changing social roles so that the original Selector was now the Recipient, neither Sarah nor Sheba, two adult females with a wealth of previous cognitive training and experience, were able to solve the problem. They simply could not inhibit selecting the larger array, despite the immediate penalty of seeing the larger amount of candy handed over to their partner.

We were astounded at their inability to learn what we saw as a simple discrimination task, given the animals' prior histories of sophisticated demonstrations with a variety of complex tasks. Clearly, a powerful mechanism was interfering and preventing them from taking advantage of their position of power within the context of the quantity judgment paradigm. Were they simply being altruistic? Both animals demonstrated great distress, both behaviorally and vocally, as soon as they selected the larger array and then watched in frustration as the collection of goodies was provided to their partner. Given our previous success with Arabic numerals within the counting tasks, we decide to try replacing the candy arrays with Arabic numerals. Now the choosing subject might encounter the numeral one versus the numeral four, instead of candy arrays. From the very first trial, the very first session, and thereafter whenever numerals were employed as stimuli, all numerically-competent chimpanzees we have tested to date have consistently selected the smaller of the two numbers. Consequently, the smaller quantity (represented by a smaller numeral) was set aside, and the chimpanzees each garnered an array of candy represented by the larger remaining numeral. Indeed, over the past five years of subsequent versions of the quantity judgment task, the chimpanzees continue to manifest a significant interference effect in optimal responding if collections of small items (either edible or non-edible) are presented for comparison. If, however, numerals are used, all the chimpanzees are successful in reliably selecting the quantity represented by the smaller numeral, and thus receive as their reward a collection of candies that corresponds to the larger remaining numeral. In addition, several other laboratories, evaluating performance on similar and/or identical tasks with several other species (pigeons, squirrel monkeys, and rhesus macaques) have all shown the same type of interference effect and failure to inhibit selection of the numerically larger array (Anderson, *et al.*, 2000; Emmerton, in press; Silberberg and Fujita, 1996). The symbolic phase of the experiment has not been replicated in another laboratory or with another species, though there is no reason to doubt that, given comparable training histories, other great apes would perform similarly.

2. CHIMPANZEES AS ARCHITECTS OF REALITY?

A second series of studies that we have been conducting over the past several years focuses on increasing our understanding of complex information-processing capacities in the nonhuman primate model, with ultimate comparisons to the documented abilities of young human children. Of particular interest is the ability of our chimpanzees to map the physical world onto multiple representational systems, such as scale models or photographs. Our pilot studies were inspired by work with 2-1/2 and 3-year-old children, and to date we have documented the chimpanzee's ability to understand the representational nature of a scale model, photographs, or videotaped images of a larger space. In addition, our ape subjects have shown us several areas of difficulty that can arise during the scale model task, likely related to attentional differences among and between chimpanzees, and that can make optimal and efficient performance, particularly by male chimpanzees, quite challenging.

2.1 Chimpanzees' Understanding of Scale Models

Recent data collected at The Ohio State University Chimpanzee Center, where we maintain a social group of seven adult and two infant chimpanzees, have indicated that chimpanzees can learn to recognize the relationship between a scale model and its larger, full-size real-world referent. We have completed a series of studies with scale models, using an approach closely modeled after DeLoache's work with young human children (e.g., DeLoache, 1987, 1991, 1995). She found that, under some test conditions, children showed an age difference in understanding how scale models compared with their full-size counterparts. DeLoache and her colleagues asked children ages 2-1/2 or 3 years to find a full-size toy hidden in a real room after the children witnessed a miniature toy being placed in the analogous hiding site within a 1:7 scale model. The older children (age 3 years) readily retrieved the toy, while the 2-1/2 year-olds had difficulty with the initial task. The diminished performance of younger children could not be accounted for by failure of the children to remember the original hiding site in the model. When asked, they were able to tell the experimenter where to find the miniature toy in the model, following their search for the actual toy in the adjacent analogous test room. DeLoache (e.g., 1995) proposed that the failure of the younger children could be based on their inability to form a "dual orientation" to the model. That is, in order to solve the scale model task, children must first be able to represent the model as a real, tangible object, and second as a symbolic representation for the actual room itself. DeLoache supported the dual representation hypothesis with results from trials during which the same younger children were shown photographs of the room, instead of the scale model. With this procedural change, both age groups were equally successful at the task. Thus, in situations in which the 2 year-olds may have had to form a dual orientation (original scale model task), their performance was poorer. Photographs, however, were readily interpreted representationally, perhaps due to young children's typical experience with picture books.

We have completed several initial experiments to investigate the ability of adult chimpanzees to understand the topographic relationship between a scale model and the full-size space represented by the model. The first study used a scale model of an indoor room that was familiar to two adult chimpanzees, Bobby and Sheba. Both animals had been involved in cognitive and behavioral training for the past 12 to 15 years, and, although quite large and fully grown, were still able to interact safely outside their home cage area with the first author. A second scale model, representing the chimpanzees' outdoor play enclosure, was used in subsequent experiments, and permitted testing of all seven adult chimpanzees in their home housing and outdoor play areas (Figure 25).

2.2   The Indoor Scale Model Study

Both Sheba (15 years old) and Bobby (10 years old) were tested in a playroom that was very familiar to them. Throughout testing, the room contained four furnishings: a large blue metal cabinet, an artificial tree, a large blue plastic tub, and a brightly-colored fabric chair. A scale model 1/7 the size of the room was placed directly outside in an adjacent hallway. The model contained miniature versions of the furniture, carpet, and other permanent features of the room. A can of soda and a miniature version of the can were used as hiding items. Both Sheba and Bobby completed eight trials that included three phases, including first, the Hiding Event, during which the subject watched as the experimenter placed the miniature can in one of the four possible hiding places in the model (Figure 26).

Figure 25. A scale model, representing the chimpanzees' play enclosure, shown here, permitted testing of all seven adult chimpanzees in their home housing and outdoor play areas.

Figure 26. During the Hiding Event, the subject watched as the experimenter placed the miniature can of soda in one of the four possible hiding places in the model.

The experimenter next showed the chimpanzee a real can of soda and then moved into the full-size room to hide it out of the subject's view. After hiding the can of soda, the experimenter returned to the hall where the chimpanzee was waiting near the model. Now the chimpanzee was allowed access to the room to search for the real soda (Retrieval 1, see Figure 27).

Figure 27. After the experimenter hid the real can of soda, the chimpanzee was allowed access to the room to search for it.

After searching in the room and finding the soda, the chimpanzee returned to the model in the hall and was encouraged to indicate where the miniature object had originally been hidden (Retrieval 2). This second retrieval served as a test of the chimpanzee's ability to remember the hiding location in the model, thus assuring that if s/he failed to find the hidden object in the real room, it was not due to memory limitations. Finally, the subject was allowed to drink the soda.

Given both chimpanzees' extensive experience on a variety of cognitive tasks and their success with one initial orientation trial with the scale model, we were surprised to find that, after completion of the formal test trials, only

Sheba was able to find the soda can in the real room at a level that was statistically above chance (Figure 28).

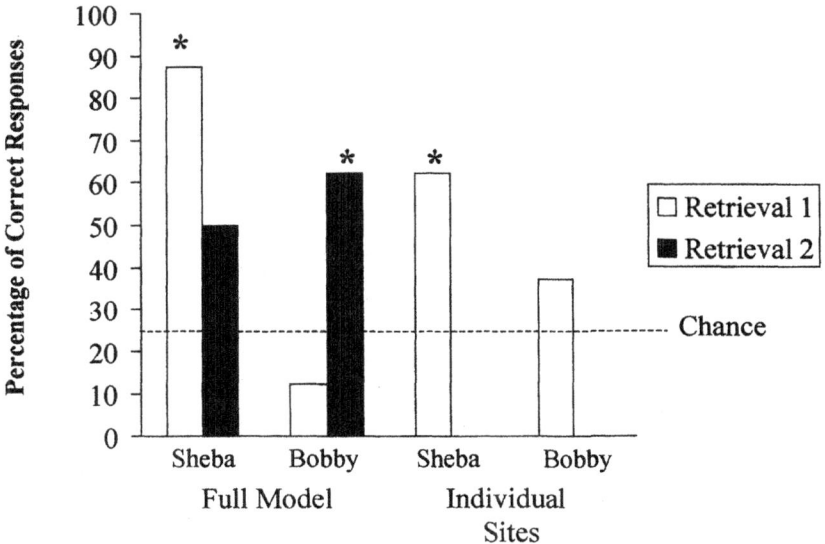

Figure 28. Though both Bobby and Sheba had extensive experience on a variety of cognitive tasks, only Sheba was able to find the soda can in the real room at a level that was statistically above chance.

Sheba typically entered the room, moved directly to the hiding site, and found the can on seven out of eight trials (each site was used twice). Her performance suggested that she was able to use the model as a source of information for determining the analogous hiding site in the actual room. Bobby, however, approached the task quite differently. Although he was very successful with Retrieval 2 (63% correct responses (CR), $p < .05$), when he had to indicate the original hiding site in the model, he was not able to efficiently locate the hidden can in the real room. On all but one trial, he entered the room and immediately searched under the blue tub, which had been randomly assigned as the correct hiding location for the first test trial. If unsuccessful, Bobby would routinely search the other sites in the room in a highly stereotyped manner, moving clockwise from site to site, until eventually finding the hidden can.

Since DeLoache's younger subjects had done better with photographs than the original scale model task, we wondered whether Sheba and Bobby could use information from color photographs of the room or the individual hiding sites to find the object that had been hidden in the room represented in the photo. Of particular interest was whether Bobby, who had failed previous tests with the scale model, might be able to perform better with photographs, and therefore show evidence of the "picture superiority" effect reported by DeLoache (1987) for her younger subjects. Test procedures with the photograph task were identical to the scale model version, with the same full-size room and hiding sites used. However, now the model was replaced with color photographs of each of the four hiding sites. The photos were mounted on the wall in the hallway outside of the test room. Again, of the two chimpanzees tested, Sheba alone found the soda can in the real room at levels above chance (63% CR, p < .05) (see Figure 28). Bobby responded in the same perseverative manner as he did with the model, although his initial preference for a specific site changed when photographs were used. Instead of the blue tub, he now chose the blue cabinet on six of the eight test trials.

Given these results, we next modified the procedures in the photograph task in two different ways to see if Bobby's performance might improve. The first change was an attempt to minimize any possible visual or attentional interference when all four photographs were visually available. A total of eight trials were conducted with only the photograph of the specific hiding site being used presented on a given trial. Bobby's performance was unchanged and similarly poor. Next, the photographs of the sites were replaced with a panorama photograph of the entire room that showed all four hiding sites simultaneously. Despite these changes, Bobby's poor performance remained unchanged, while Sheba continued to be successful no matter what modality was used for stimuli (individual site photos: 63% CR, p < .05; panorama photo: 75% CR, p < .05).

A final modification in our approach was to present a pre-recorded video representation of the Hiding Event to the chimpanzees, with the experimenter using dramatic animation as she hid the can, including gestures and enthusiastic vocal communication in an effort to draw the chimpanzees' attention to the video vignette of the Hiding Event. We hypothesized that this modality, while having the two-dimensionality of photographs, might convey more information with the added elements of sound and action, and therefore might help enhance Bobby's continued performance. With this approach, the animals were presented with a brief (10s) video presentation of the experimenter hiding the real can of soda in the full-size room. Following their viewing of the video sequence, the subject was again allowed access to the room to find the soda. Sheba, as before, performed successfully with this shift in modality, and continued to locate the hidden can at levels above chance (5/8 CR, p < .05). Bobby was again unable to efficiently search the room, and exhibited similar perserverative choice patterns, moving from site to site in a clockwise search pattern.

The indoor scale model, photograph, and video tasks demonstrated that a chimpanzee could recognize the relationship between several types of spatial representations and the larger space they represented. The data also suggested that one subject, Sheba, was not only able to understand that the room and model were related, but likely understood that the elements of one were analogous to elements in the other. Events that occurred in one represented events that could occur in the other. It is unclear, however, whether Bobby's poor performance was due to an inability to form the type of dual orientation with the model that DeLoache (e.g., 1995) had suggested was necessary in order for children to be successful with the scale model comparisons, or because of an inability he may have had to inhibit a rigid, perseverative search pattern. Alternatively, it may have been some combination of both. If Bobby's difficulty was an inability to represent the model as both object and symbol, as suggested for young children, his performance should have improved when the photographs and especially videotaped scenarios (which would not have been perceived as objects) were used. However, instead of showing a "picture superiority" effect, Bobby continued to show inefficient and incorrect search behaviors.

2.3   Addressing Individual or Sex Differences: The Outdoor Scale Model Study

In order to test all seven of the adult chimpanzees housed at the Chimpanzee Center, a 1:7 scale model of one section of their outdoor enclosure was constructed (Figure 29).

All but two of the animals (Digger and Abby) had extensive experience on other cognitive tasks (e.g., Boysen, *et al.*, 1996; Limongelli, *et al.*, 1995; Thompson, *et al.*, 1997). The outdoor scale model task was similar to the previous indoor scale model study, with both the new model and actual outdoor enclosure used for tests containing four hiding sites marked by large toys that were familiar to all of the chimpanzees. The toys were arranged in four positions (see Figure 30) and remained stationary for all trials. Plastic bottles filled with fruit juice and a miniature replica bottle were used as hiding items.

Figure 29. A 1:7 scale model of one section of the chimpanzees' outdoor enclosure was constructed.

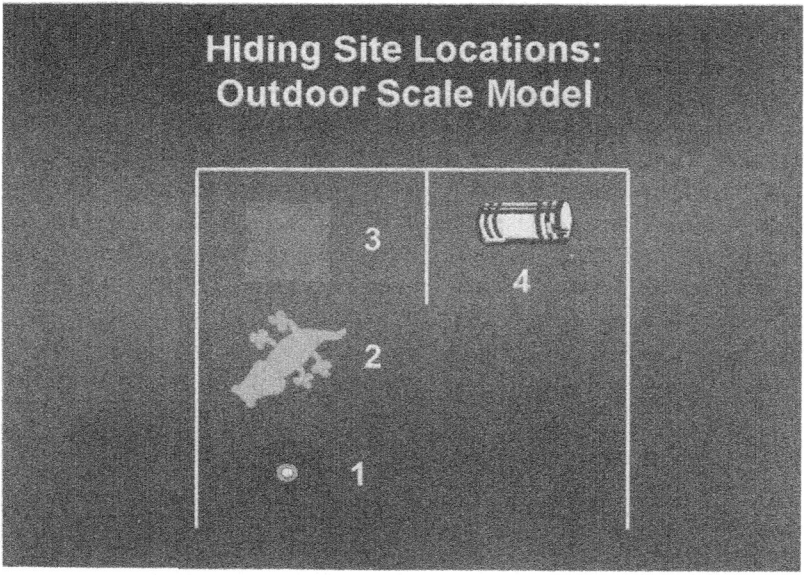

Figure 30. The toys were arranged in four positions and remained stationary for all trials.

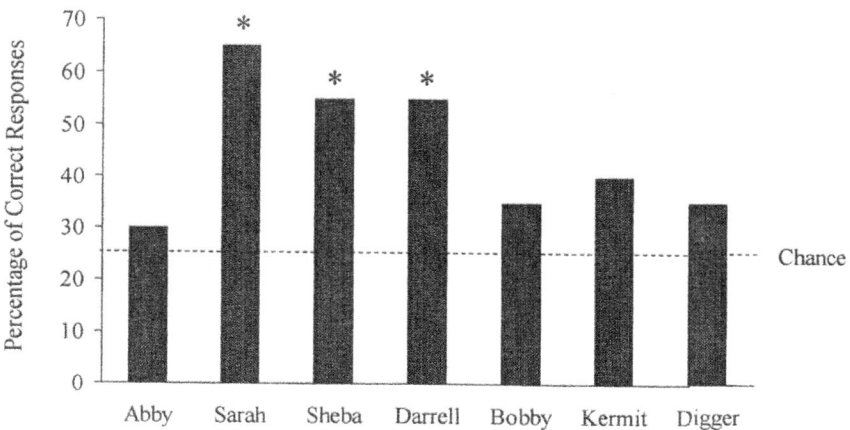

Figure 31. All seven chimpanzees had experience with cognitive tasks.

Three of the seven chimpanzees tested performed at levels significantly above chance (Sheba: 11/20 CR, p < .05; Sarah: 13/20 CR, p < .05; Darrell: 11/20 CR, p < .05) (Figure 31). The results replicated Sheba's performance with the indoor model and demonstrated that other chimpanzees were able to understand the relationship between a model and its referent. Analysis of the choice patterns of the unsuccessful subjects was also interesting. All four animals showed a significant preference to visit the first test site, which was a small tire from a garden tractor, and they visited this site on each trial, regardless of whether it was correct. Second site choices were also analyzed. One subject, Abby, chose the correct site in her second search attempt significantly above chance (12/14 CR, p < .05). She would first search the tire in Position 1, then move directly to the correct site and retrieve the juice. The other three subjects who failed to perform adequately did not use the same strategy. This included Bobby, a young adult male, Digger, an adolescent male, and Darrell, an adult male. Instead, after their initial incorrect choice of Position 1, the three males frequently visited the adjacent site at Position 2 (Figure 32).

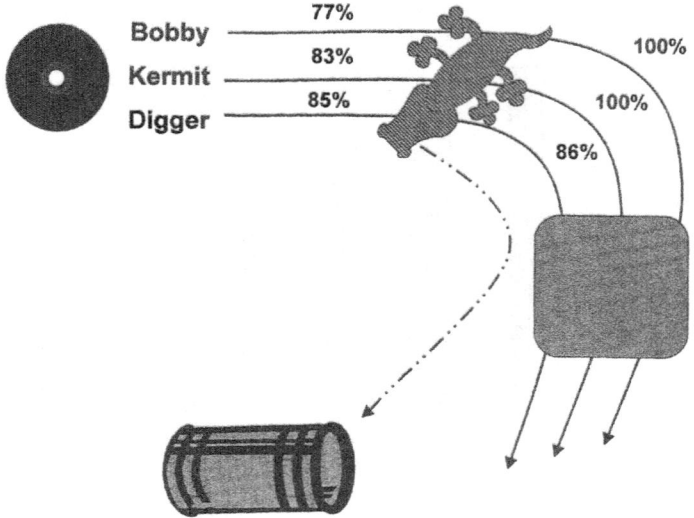

Figure 32. After their initial incorrect choice of Position 1, the three males frequently visited the adjacent site at Position 2.

If third choices were then made, they were sites that were in the same general direction (Positions 3 or 4). These chimpanzees appeared to be using a search method that involved checking each site, moving in a clockwise manner around the enclosure until they located the hidden juice bottle.

In an effort to further understand the difficulties that these four subjects were having with the task, we attempted to break up the rigid search patterns they were using by moving the items in both the test enclosure and the model to new positions on every trial. With this change, the chimpanzees encountered a different toy in Position 1, etc., on each trial. Following testing, both Sarah and Sheba were again successful, as was the third adult female, Abby, who performed at the same level as Sheba and Sarah (for all: 5/8 CR, $p < .05$). Abby's success was particularly interesting given her performance when the sites were fixed. Moving the sites did seem to break her "routine" of visiting Position 1 before going to the correct site as her second choice. Abby's performance was also particularly striking given that she has had no prior cognitive training or testing experience, has retinal damage due to diabetes, and was a former pet who lived, isolated from other chimpanzees, in a human household for the first 20 years of her life.

The other four chimpanzees tested, all males, did not perform at levels that reached statistical significance. Darrell, whose performance had been comparable to Sheba's and Sarah's in the initial phase of testing, showed some deterioration with the procedural change (moving the sites each trial), and thus his score fell below statistical significance. However, both older adult males (Darrell and Kermit, age 20 years) chose the correct site on 50% of trials. The two younger males, Digger and Bobby, were again unsuccessful and had the poorest performance of the seven animals. When we examined what the unsuccessful chimpanzees were doing as they entered the enclosure, again we found a preference for Position 1, regardless of the item located there. The unsuccessful subjects also demonstrated a high rate of visiting the adjacent site for their next search, and then continued looking at the sites that were positioned in the same, clockwise direction (Figure 33). Thus, with the exception of one subject (Abby), moving the sites between trials did not attenuate the rigid search patterns used by the unsuccessful subjects.

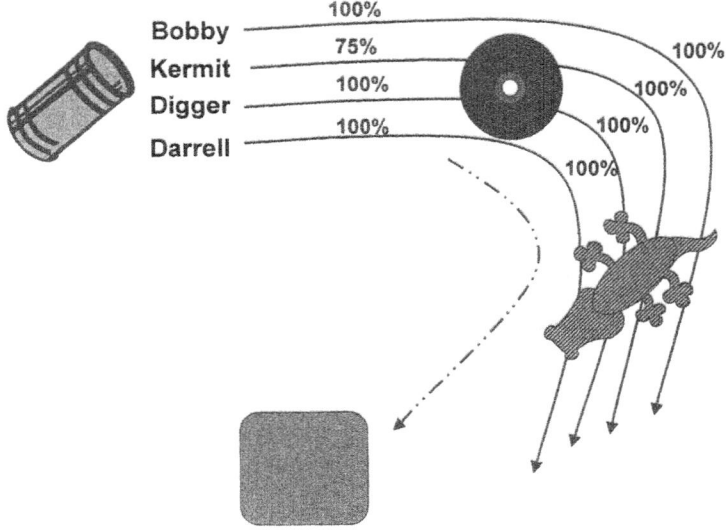

Figure 33. The unsuccessful subjects also demonstrated a high rate of visiting the adjacent site for their next search, and then continued looking at the sites that were positioned in the same, clockwise direction.

3.   SUMMARY AND CONCLUSIONS

The results from these preliminary studies show that chimpanzees are capable of understanding the relationship between a scale model or photographs of the corresponding real space. The results replicate and extend the innovative studies of similar skills in children (e.g., DeLoache,

1987, 1991) to a nonverbal species with whom previous attempts to demonstrate such understanding had failed (Premack and Premack, 1983). However, recognizing these relationships was not necessarily demonstrable in all chimpanzees, as evidenced by the failure of some of the subjects to successfully complete the task. The second study, using a model of the animals' outdoor enclosure, allowed us to examine the error patterns shown by unsuccessful subjects. Such search patterns differed from those reported for young children (2-1/2 years). The younger children were similarly unable to complete DeLoache's scale model task but did not show the perseverative search patterns used by the chimpanzees. It may not have been an inability to form a dual orientation that plagued the unsuccessful ape subjects, but rather in the inability to inhibit patterns of deeply rooted behavior. Nonetheless, specified variables that contribute to forming such complex representations as tested in these experiments have not been adequately defined or empirically tested for either human children or for chimpanzees.

REFERENCES

Anderson, J., Awazu, S., and Fujita, K., 2000, Squirrel monkeys (*Saimiri sciureus*) learn self-control in food array selection task with a reverse reward contingency, *J. Experimental Psychology: Animal Behavior Processes* 26(1): 87-97.

Barrow, J., 1992, *Pi in the Sky*, New York: Academic Press.

Boysen, S.T., 1993, Counting in chimpanzees: Nonhuman principles and emergent properties of number. Pp. 39-59 in: (Eds. S.T. Boysen and E. J. Capaldi), *The Development of Numerical Competence: Animal and Human Models*, Hillsdale, NJ: Lawrence Erlbaum Associates.

Boysen, S.T. and Berntson, G.G., 1989, Numerical competence in a chimpanzee (*Pan troglodytes*), *J. of Comp. Psych.* 103: 23-31.

Boysen, S.T. and Capaldi, E.J. (Eds.), 1993, *The Development of Numerical Competence: Animal and Human Models*, Hillsdale, NJ: Lawrence Erlbaum Associates.

Boysen, S.T., Berntson, G.G., Hannan, M.B., and Cacioppo, J.T., 1996, Quantity-based interference and symbolic representations in chimpanzees (*Pan troglodytes*), *J. of Experimental Psychology: Animal Behavior Processes*. 22: 76-86.

Davis, H. and Perusse, R., 1988, Numerical competence in animals: Definitional issues, current evidence, and a new research agenda, *Behavioral and Brain Sciences* 11: 561-579.

DeLoache, J.S., 1987, Rapid change in the symbolic functioning of very young children, *Science* 238: 1556-1557.

DeLoache, J.S., 1991, Symbolic functioning in very young children: Understanding of pictures and models, *Child Development* 62: 736-752.

DeLoache, J.S., 1995, Early symbol understanding and use. Pp. 65-114 in: (Ed. D.L. Medin), *The Psychology of Learning and Motivation*, San Diego, CA: Academic Press.

Emmerton, J., in press, Biases in pigeons' responses to numerosity: Discrimination of "more" versus "less," *J. Experimental Psychology: Animal Behavior Processes*.

Gallistel, C.R., 1989, Animal cognition: The representation of space, time and number. Pp. 155-189 in: (Eds. M.R. Rosenzweig and L.W. Porter), *Annual Review of Psychology*.

Gallistel, C.R., 1990, *The Organization of Learning*. Cambridge, MA: MIT Press.

Limongelli, L., Boysen S.T., and Visalberghi, E., 1995, Comprehension of cause-effect relations in a tool-using task by chimpanzees (*Pan troglodytes*), *J. of Comp. Psych.* 109: 18-26.

Meck, W.H. and Church, R.M., 1984, The numerical attribute of stimuli. In: (Eds. H.L. Roitblat, L. Herman, and H.S. Terrace0, *Animal Cognition: Proceedings of the Harry Frank Guggenheim Conference, June 2-4, 1982*. Hillsdale, NJ: L. Erlbaum Associates.

Premack, D. and Premack, A.J., 1983, *The Mind of an Ape*, New York: W.W. Norton and Company.

Silberberg, A. and Fujita, K., 1996, Pointing at smaller food amounts in an analogue of Boysen and Berntson's 1995, procedures, *J. Exper. Anal. Behav.* 66: 143-147.

Thompson, R.K.R., Oden, D.L., and Boysen, S.T., 1997, Language-naive chimpanzees (*Pan troglodytes*) judge relations-between-relations in an abstract matching task, *J. of Experimental Psychology: Animal Behavior Processes*, 23: 31-43.

SECTION FOUR
GORILLAS, THE GREATEST OF THE APES
INTRODUCTION

Few primate species capture the public imagination more than does the largest living primate, the gorilla. Some taxonomists recognize two gorilla species (*Gorilla beringei* and *G. gorilla*) and four or five subspecies (*G. beringei graueri, G. b. ssp?, G. b. beringei, G. gorilla diehli, G. g. gorilla*) (IUCN *Red List of Threatened Species*, 2000), while others more traditionally recognize one species (*Gorilla gorilla*) and three subspecies (*G. g. beringei, G. g. gorilla, G. g. graueri*). Three of the five subspecies are classified as critically endangered by IUCN specialists; the remaining two subspecies are endangered (IUCN *Red List of Threatened Species*, 2000). By any estimate and however they are classified, the survival outlook for gorillas is bleak. Wild gorillas are threatened by habitat loss and are found in increasingly fragmented and isolated populations. Some subspecies, such as the mountain gorilla (*G. b. beringei*) and the Cross River gorilla (*G. g. diehli*), number fewer than 325 individuals. Many gorilla populations are distributed in countries plagued by recurrent civil unrest (Stanford, 1999). More recently, gorilla populations have been decimated by the ever-growing bushmeat trade, through which hundreds or thousands of apes are killed each year (Stanford, 1999; see also Rose, this volume). The trade in bushmeat fuels the pet trade, a now-thriving market based on infant gorillas orphaned through bushmeat hunting (McRae, 2000).

Francine Patterson and Marilyn Matevia open section four of this volume with an overview of the status of wild gorillas. They describe a truly dire situation, but they believe that the interest centered on the captive gorilla Koko, famed for language and other tests of cognition, provides some hope for the survival of her wild counterparts. Patterson and Matevia note the public's profound interest in Koko's thoughts and feelings. This interest and appeal is not limited to the west—the distribution of the children's book *Koko's Kitten* is beginning to shape views of gorillas in African countries as well (see also Rose, this volume). An understanding of gorillas' mental capacities, which are further summarized in this section by Francine Patterson and Wendy Gordon, facilitate the human notion of connection to gorillas. Ultimately, this sense of kinship may form our best hope for the gorilla's survival.

Captive husbandry techniques and management policies for the gorilla and for other species have changed dramatically over the past several decades (Ogden and Wharton, 1997). Patterson and Matevia summarize what has been learned by gorilla keepers and how this information, coupled with increased understanding of gorilla communication with human caretakers, can be used to improve captive gorillas' mental and physical health.

Despite the pioneering research of Dian Fossey (Fossey, 1983) and other, more recent studies of wild gorillas, gaps remain in our knowledge of intragroup dynamics and how new gorilla groups are formed. Captive research enables a more detailed account of within-group power shifts, as described by Tamara Stein in her analysis of the ascendancy of a blackback male as the leader of a group of western lowland gorillas (*G. g. gorilla*) housed at the Brookfield Zoo (Brookfield, IL). In the wild, few cases of group formation or transitions of power have been reported. Stein describes how group members sought to maintain close proximity to the blackback male and increasingly treated him as the group leader. Eventually, he mated with a female even though the silverback remained in the group. This observation suggests that under some circumstances a blackback male can assume the leader role at earlier ages than is common in the wild. Whether this power shift occurs without strife, as reported by Stein, remains to be seen and has relevance for how captive groups are managed. Stein's observations are particularly intriguing in light of Robbins' (1999) account of subordinate males mating while living in multimale groups of wild mountain gorillas (*G. b. beringei*). Clearly, many hypotheses about gorilla group formation and maintenance remain to be explored.

REFERENCES

Fossey, D., 1983, *Gorillas in the Mist*, Boston, MA: Houghton Miflin.
IUCN, 2000, *Red List of Threatened Species*. Eletronic document, http://www.iucn.org, accessed January 2001.
McRae, M., 2000, Central Africa's orphan gorillas, *National Geographic* 197(2): 84-97.
Ogden, J. and Wharton, D. (Eds.), 1997, *Management of Gorillas in Captivity*, The Gorilla Species Survival Plan and the Atlanta/Fulton Co. Zoo.
Robbins, M.M., 1999, Male mating patterns in wild multimale mountain gorilla groups, *Animal Behaviour* 57(5): 1013-1020.
Rose, A.L., this volume.
Stanford, C.B., 1999, Gorilla warfare, *The Sciences* 39(4): 18-28.

Chapter 10

THE STATUS OF GORILLAS WORLDWIDE

F.G.P. Patterson and M.L. Matevia
The Gorilla Foundation/koko.org, P.O. Box 620530, Woodside, CA 94062

1. INTRODUCTION

We describe here the status of free-living gorillas in terms of their distribution, abundance, and the threats to their continued existence. We also address a variety of aspects of captive management of gorillas. We report both good and bad news about current field research and conservation work, and we describe a few *in-situ* and *ex-situ* sanctuary scenarios. We conclude with a perspective on public education and awareness raising: the urgent need for creating what really amounts to a mass social movement to save the great apes (Rose, this volume).

2. A BRIEF OVERVIEW OF PROJECT KOKO

Originally, we started the interspecies communication study, now overseen and supported by The Gorilla Foundation, as a dissertation project. In the 29 years since, however, it evolved into much more. After many years of working especially closely with two members of the lowland gorilla species, our perspective has become much more global. It is not simply to learn all we can about an individual—it is to save a species. We started working just with Koko, and then we learned that we must protect all gorillas because they each have intrinsic worth.

Our goals expanded from the study of the cognitive development of Koko and Michael to the creation of a sanctuary for them and others like them in captivity, one that is appropriate for their needs as shaped by what we have learned. We have also begun preliminary investigations into the opportunities for co-sponsoring African reserves. We need to set aside habitat for free-living gorillas before it is too late. At the same time, and every bit as essential, we need to heighten awareness: people must know that gorillas are our kin, that our kin are dying at our hands, and that they are worth saving. There is among our own species an almost universal fascination with gorillas. In the parlance of zoo officials and planners, gorillas are considered "charismatic megafauna" (attention-getting big animals). If we can transform that fascination into conservation, we may have a chance of saving the species.

3.   DISTRIBUTION AND POPULATIONS OF GORILLAS

There are four or five subspecies of gorilla: the mountain gorilla (*Gorilla beringei beringei*), the Bwindi gorilla (*G. b. beringei*), Grauer's gorilla (*G. b. graueri*), the Cross River gorilla (*G. gorilla diehli*), and the western lowland gorilla (*G. g. gorilla*) (IUCN *Red List of Threatened Species*, 2000). All of these populations, free-living members of which exist only in Africa, have been in decline. In fact, the 2000 World Conservation Union (IUCN) *Red List of Threatened Species* classified both species as "endangered," with three subspecies listed as "critically endangered."

There are two primary reasons for this decline. One is a steady elimination of habitat. The range over which these subspecies exist is approximately 311,000 km^2. The actual space they occupy is about one-tenth of that, only 31,000 km^2, or an area smaller than the state of Ohio.

Remarkably, there are some small isolated populations, possibly subspecies, in the eastern side that have yet to be explored. In the west, the isolated subspecies the Cross River gorilla (classified by the IUCN as "critically endangered"), which exists in Cameroon and Nigeria, is cut off by about 200 km from other gorilla populations. Genetic studies of mitochondrial differentiation suggests that they may be as different from western lowland gorillas as chimpanzees are from bonobos (Morell, 1994).

The space left for mountain gorillas is about 375 km^2—an area only slightly larger than that covered by the Kuala Lumpur Airport, where conference participants transited in Malaysia. The remaining territory for Bwindi and Grauer's gorillas is approximately 15,000 km^2 and that of western lowland gorilla is approximately 50,000 km^2. Less than 10% of the western gorilla population resides within protected parks (Oates, 1996). Historically, biologists viewed expanding villages and farms as the greatest threat to gorilla habitat. The last decade has seen decimation on a far greater scale, because foreign timber companies have been drawn to the inexpensive and largely unregulated logging opportunities in west African rainforests.

Populations of the five gorilla subspecies are declining as dramatically as their habitats. Western lowland gorillas (*G. gorilla gorilla*) are indigenous to seven countries: Angola, Cameroon, Central African Republic, Congo, the Democratic Republic of the Congo, Equatorial Guinea, and Gabon. The most recent population estimates for this subspecies range from 92,000 to 110,000, with the greatest concentrations in the Congo (44,000) and Gabon (43,000). However these numbers might already overestimate true population sizes, because western lowland gorillas are being hunted intensively in the burgeoning bushmeat trade. The Cross River gorilla (*G. g. diehli*) numbers only 150-200 individuals in Cameroon and Nigeria. There are approximately 8,000-17,000 Grauer's gorillas (*G. beringei graueri*), all in the Democratic Republic of the Congo. Of the mountain gorillas (*G. b. beringei*), now critically endangered and grievously close to extinction, there are at most 350 existing in the Democratic Republic of the Congo, Rwanda, and Uganda. The Bwindi gorilla (*G. b. ssp.?*) numbers less than 250 mature individuals and is found in the Impenetrable Forest of the Democratic Republic of the Congo and Uganda (all population estimates are from the Harcourt, 1996; IUCN *Red List of Threatened Species*, 2000; Oates, 1996).

4.   GORILLAS IN CAPTIVITY

Worldwide, about 700 gorillas live in captivity, 355 of them in facilities in the United States. Almost all of these captive gorillas are the western lowland subspecies. Husbandry has been increasingly successful. At the April 1998 meeting of members of the Western Lowland Gorilla Species Survival Plan (SSP), it was reported that during the previous year, births exceeded deaths 14 (7 males and 7 females) to 10 (6 males and 4 females). Recently, a husbandry manual was prepared through a cooperative of institutions with gorillas in their care (Ogden and Wharton, 1997).

4.1   Health and Well-being

There is still much to learn about the health and well-being of gorillas. Cardiovascular disease causes 41% of all mortality in adult captive gorillas (Meehan, 1997). Notably, the majority of these deaths are caused by fibrosing cardiomyopathy and aortic dissection, not myocardial infarction. In fibrosing cardiomyopathy, the wall of the heart muscle grows abnormally thick and stiff. Aortic dissections are essentially aneurysms, in which the aorta swells and bursts. Death of gorillas by these cardiac events is usually sudden and often without prior clinical evidence of disease. Cardiomyopathy may be caused by genetic factors, viral infection, hypertension, arteriosclerosis, obesity, stress, and infectious or toxic agents (Schulman, *et al.*, 1995). The major risk factor for aortic dissection in humans is hypertension (Kenny, *et al.*, 1994), but we have no blood pressure studies of unanesthetized gorillas for comparison.

Professionals who care for gorillas speculate that, as with humans, improper diet and insufficient exercise make gorillas particularly susceptible to these diseases. Mortality rates of male gorillas exceed those of female gorillas. We hypothesize that because of the nature of gorilla social organization, captivity is especially stressful for males. This, in turn, makes them more vulnerable to these unusual cardiac events. Male gorillas are responsible for their group, and if they were free-living, they would move the group away from humans. They cannot do this in a captive setting. At the Gorilla Foundation, we work in a situation where the gorillas in our care are not observed by thousands of people everyday. When strangers come, mainly to repair things, we find that the males become highly stressed. They have diarrhea, they shake, they cry, and they sweat. They sometimes do not recover from these symptoms for days after exposure to strangers. We have noticed similar symptoms in male gorillas in films we have seen of "habituated" free-living groups and have confirmed our observations with one field researcher (T. Butynski, pers. comm.). Some wild silverbacks have diarrhea when they are stressed and they are upset—they are very uncomfortable with having to keep their groups in proximity to humans.

4.2   Gorilla Temperament

There are some characteristics of gorilla temperament that may contribute to this sensitivity. Since Robert Yerkes first wrote extensively about them in the 1920s, gorillas have had a reputation for being stubborn, aloof, and even stupid. They are definitely stubborn, they can certainly be aloof, but they are far from stupid. They simply have a brute strength and iron will that makes them poor subjects for the tests of problem-solving and cleverness that have historically been presented to their more diminutive and dexterous relatives, the chimpanzees.

Within their patriarchal social structure are very close-knit family groups. Once a male has established a group with several females, he typically leads and protects that group for as long as he lives. The whole group is, in turn, extraordinarily protective of its young.

The ghastly bushmeat trade has brought to prominence another characteristic of gorilla temperament: they are psychologically fragile. When the adults of a group are killed for sale as bushmeat, the infants sometimes end up in villages to be "fattened up" for later consumption or to be sold as pets. They almost always die first. Apparently they simply cannot withstand the loss of their protective and close-knit social group. Bonobos are similarly fragile. Chimpanzees in such circumstances have a much higher likelihood of survival.

Gorillas, in our experience, are also extremely observer-sensitive, and this relates again to the stress they experience in captive settings. Until Koko did so in 1982 (Patterson and Cohn, 1994), no gorilla had passed the classic mirror test of self-recognition. Perhaps they failed not because they lack self-awareness, but because of the gorilla's heightened self-consciousness about being stared at, which could be triggered by looking at themselves in mirrors, not to mention by scientists observing their behaviors after a marking test. When Michael was given a videotaped marking pretest, he looked at himself in the mirror, then retreated to the back of the room, turned his back, and wiped his face.

Gorillas go to great lengths to maintain control over their environment. There are many anecdotal reports of gorillas being "control freaks". Examples of a gorilla's clear comprehension of a request or command as exhibited by their eventual compliance, but executed (in what gorilla caregivers sometimes refer to as "gorilla time") long moments later. We recorded this in our early journals of interactions with Koko. In one case Koko was asked to retrieve a particular object from her room. She brought everything *but* the specified object, which she actually seemed to deliberately avoid.

Gorillas are highly expressive, making extensive use of a natural system of postures, gestures, and facial expressions. Indeed, this is what makes Koko and Michael such good students of sign language. The predisposition was there, we merely expanded their repertoire. Joanne Tanner (1998) conducted a study of gorillas residing in zoos and documented 60 different gestures that are very similar or identical to sign language gestures. Their gesture use is subtle, spontaneous, and creative. Gorillas also have a rich repetoire of communicative vocalizations. The assent grunt sounds like "uh-huh" and means "yes," and the bark sounds like "unh, unh" and means "no." Only during the last two decades have the scientific and zoo communities observed and acknowledged these extensive communicative systems.

4.3   Designing Appropriate Captive Environments

Given what we now know about these mental and emotional characteristics of gorillas, captive conditions raise a number of concerns.

4.3.1   Position Relative to Visitors

In the classic gorilla exhibit of the 1950s, '60s and '70s, gorillas were always at or beneath the human visitor's eye level. Gorillas are very uncomfortable with this kind of arrangement. They need to feel safe and that they can easily survey their surroundings, and a gorilla does not feel secure with humans standing above him or her. After Jane Goodall visited us in the 1970s, she wrote to ask Koko's input for her field observers, who did not want to crouch when they observed the chimpanzees. She asked Koko, "do you like to be watched by people who are standing up or sitting down?" We relayed the question to Koko, who answered "down." We asked her again, to be sure, and she repeated "down." We asked her one last time, and she said "sleep," and then she laid down to make the point completely clear. She likes people to reduce their stature and be flat, below her. When Michael takes a drink from a caregiver outdoors, he brings a large section of PVC piping, which he stands on end and sits on, so that his head is higher than the caregiver's.

The most recent and more sensitive zoo exhibits have taken this need into account. Even in the indoor work areas, the gorillas' holding areas are elevated, so that caregivers move about at a level approximately 1m below the gorillas. The gorillas seem to feel more comfortable with their caregivers in this arrangement.

4.3.2   Flight Space

Another important factor in exhibit design is "flight space." Gorillas need to be able to hide, or at least retreat, from human observation when they are on display. Some designers still do not build these areas, because of concern that the gorillas will never be in view of the public. In actuality, the availability of flight space makes a gorilla's day on exhibit much more enjoyable and far less stressful. It is important for all great apes, but especially for gorillas.

4.3.3   Control of Environment

Most obviously, the ability to communicate desires via gestures and vocalizations gives gorillas a measure of control over their generally human-controlled environment. Within an exhibit, methods can be devised for satisfying gorillas' control needs. At the Gorilla Foundation, we are consulting with a local technology company that plans to create an enrichment device in the form of a camera/software integration system that will fit the image of the gorilla being filmed into a virtual environment depicted on a monitor. The gorilla will be able to watch him- or herself in that virtual background. We hope that this technology can eventually be expanded to allow the gorillas to control such environmental factors as lighting, music, television, and certain gates in their units.

4.3.4 Diet

Diet is one of the more easily and effectively manipulated environmental factors. Free-living gorillas spend most of their day gathering and eating food. In captivity, a meal presented unimaginatively can be consumed in a few minutes, leaving many hours to fill before the next meal. During the last two decades, gorilla caregivers have found ways to make food take longer to eat and have recognized the importance of providing low-calorie (or high effort) forage and browse items between meals. In addition, caregivers at the Basel Zoo have successfully eliminated food-related aggression and regurgitation by removing fruit from their gorillas' diet. More research needs to be done on the association between eating fruits, regurgitation, and aggression at feeding. But regurgitation-reingestion habits, which seem to be strongly linked to boredom (Akers and Schildkraut, 1985), remain notoriously hard to break (Gould and Bres, 1986). We need to continue to focus creative efforts on reducing boredom.

In order to improve captive diets, more specific and thorough analyses of the dietary elements sought out by free-living gorillas are needed. Many of our assumptions and hypotheses are based on what we know about human and chimpanzee dietary needs. Gorillas have a protein requirement, which free-living gorillas meet by eating insects. Most insects are abundant on the plants the gorillas consume, but gorillas will also search for bugs (Nishihara, 1996). As with humans, gorillas do seem to have some trace element requirements. For example, 60 to 70% of captive gorillas exhibit abdominal adhesions, which some investigators have suggested may result from a copper deficiency (verbal report of the Veterinary Advisory Committee of the Gorilla SSP, April, 1998). This aspect of diet needs much more study.

4.3.5 Interactions with Caregivers

As more zoo workers come to terms with the understanding that gestural communication is natural for gorillas, we believe they will see that it can be used to facilitate husbandry routines. A common communication system will help the caregivers better understand the gorillas' needs and will enable the workers to better convey their own wishes. Every zookeeper knows that gorillas can eventually understand English commands, but a shared system will give both the caregivers and the gorillas the very rewarding experience of two-way communication.

5.   AN *EX-SITU* PRESERVE SCENARIO: MAUI, HAWAII

One of the ways that the Gorilla Foundation hopes to implement some of the issues discussed above is to create a preserve in an environment that is, as much as possible, similar to the environment in which free-living lowland gorillas reside. Seventy acres of land on Maui has been made available by one of our board members. There we hope to create the space these gorillas need and deserve. We have the potential to create social groups and to give them privacy. This facility will not be open to the public for visitation, but will be open via a satellite linkup for live video feeds and even videoconferencing. People will be able to see the gorillas from an educational center in a nearby town and from other remote viewing stations. In turn, the gorillas will be able to watch the people, if they so choose. This preserve will be used as a retirement center for our own and other gorillas, as a safe haven for gorillas who are not particularly adapted to the captive situation, and as a facility for the establishment of bachelor groups of male gorillas. We have the luxury of incorporating many years of worldwide captive management experience into this facility, but we also have the unique opportunity to ask our language-using gorillas for their opinions and desires concerning the design features. Not surprisingly, the gorillas have opinions on what they do and do not like.

6.   FIELD RESEARCH AND CONSERVATION EFFORTS

There is some good news to report from the field. There are more sites for research on free-living gorillas than ever before and more *in situ* bases for conservation work (Table 17). This development will greatly improve Population and Habitat Viability Analyses (PHVA). For example, in 1998, the Captive Breeding Specialist Group (CBSG) PHVA workshop for mountain gorillas was held in Uganda and brought together a variety of citizens in positions to influence land use decisions.

Still, there are many reasons to be alarmed. The income generated for parks and protection by the ecotourism trade is shrinking dramatically, and business dwindles due to political upheavals and terrorism.

The clash of human and gorilla interests is growing. Primates are hunted in 27 of 44 study areas and conservation projects (Oates, 1996). In May 1997, ten habituated gorillas were shot and killed in the Democratic Republic of the Congo by soldiers in the Virunga National Park. The Volcanoes National Park has been closed to researchers since fall 1994 due to continued civil strife, but has since re-opened in 1999.

Table 17. Selected gorilla field study sites

Country	Investigator(s)
Democratic Republic of Congo	E. Sarniento and T. Butynski
Uganda	C. Stanford and M. Goldsmith
Rwanda	D. Steklis, S. Madry, and L. Williamson, IGCP
Central African Republic	D. Doran, M. Remis, and A. Blom
Gabon	Station d'Etudes des Gorillas et Chimpanzees, ECOFAC
Nigeria	J. Oates
People's Republic of Congo	R. Parnell, D. Morgan, C. Olejniczak, M. Fay, and S. Blake

The combination of human agriculture and logging is dramatically shrinking the amount of forest available to gorillas in the countries to which they are indigenous. This promises to worsen. The human population growth is now estimated to triple by the year 2050 despite the death toll taken by AIDs. One third of the African population is already malnourished. Agricultural land needs will skyrocket.

During the decade of the 1980s, timber exportation increased fourfold in the 11 countries that have indigenous gorilla populations. According to some reports, rainforests are disappearing from Africa faster than on any other continent (see also Boysen and Butynski, this volume). The timber business has contributed to a flourishing bushmeat trade. Shockingly, the bushmeat trade now appears to be a more significant threat to gorilla survival than is the loss of habitat. In contrast, there are parts of the gorillas' habitats that have not yet been touched by loggers. It is important that these be protected.

6.1   The Potential for *In Situ* Preserves

The Likuala River in the northern region of the People's Republic of Congo winds through wet, swampy land. About a decade ago, to the great surprise of many researchers, large numbers of gorillas were surveyed in this territory. Estimates are that many thousands of gorillas may live in this region.

Lake Tele, in this Likouala Region about 60 km above the equator, has been proposed as a site for a preserve. The preserve area could be as large as 1,500 km^2. The land is as yet undesirable to humans; huge swaths of it are swamp. It cannot be developed economically, and draining the swamps is not yet a viable option. This area should be set aside and protected before developers begin to see it as a last resort. The goal behind this preserve should not be to open it to visitors, but to let the gorillas roam free and protected. The site should be opened to field researchers, local conservation education programs could be developed around it, and remote viewing access stations and "virtual" tours could be established.

6.2   The Bushmeat Trade

The bushmeat trade now represents the greatest threat to the continued existence of gorillas, chimpanzees, and bonobos (Dupain and Van Elsacker, this volume; Rose, this volume). In some areas apes have been hunted for food, but historically they were taken at a sustainable rate. Today, observers estimate that 3,000 to 6,000 apes are killed annually, which is not sustainable for survival of the species. The acceleration of commercial logging in west African forests has several devastating ramifications: roads are being constructed that make previously impenetrable forests more accessible to hunters; the hunters can now drive large vehicles into the forests, which allows them to kill and remove many more animals; and the logging companies purchase the bushmeat to feed their crews, increasing demand for the killing. It is illegal to hunt apes in these countries, but no one enforces the laws. The Gorilla Foundation has funded the opening of a sanctuary for gorilla orphans in Mefou Naitonal Park, Cameroon.

The hunt for gorillas is particularly brutal. Adults rush to protect the young, putting themselves in the way of hunters' bullets. Orphaned infants are typically dropped off in villages where they may endure neglect and harassment. Few survive; those who do have few options. They should be rehabilitated for life in the forest, and few such facilities exist.

7.   THE PERSONHOOD OF GORILLAS

The gorilla traits we have described comprise, in great part, what may be thought of as personhood: self-awareness, the ability to understand and use language, the ability to empathize. Koko and Michael have created representational art in which they depict emotions, events (such as an earthquake), or cherished friends or pets (Koko painted her bird, Michael painted his dog), and they use language to name those pictures. They use and construct tools. They make-believe, as when Koko acts out fighting or biting scenes with her dolls and alligators, or pretends to nurse a monkey doll or make her ape doll sign. Sometimes we are asked to play along with these scenarios, "drinking" from empty cups, or reacting as if bitten when Koko threatens us with a toy alligator. She seemingly expresses grief, pride, shame, embarrassment, humor, and deception. She appears to attribute mental states to others. Collectively, these features create a strong case for gorilla personhood (Singer and Cavalieri, this volume).

8.   GETTING THE MESSAGE OUT

During the last two years, the Gorilla Foundation (along with other organizations concerned with great apes) has made a concerted effort to expand its web presence. The web is an effective and inexpensive way to reach a growing body of people, and we use the site primarily to educate people. From time to time, the media spots something interesting on our site, writes about it, and generates even more traffic. In April of 1998, America On Line (AOL) offered to host an online chat with Koko, the first "interspecies" chat of this very popular web phenomenon. We greatly underestimated the interest in this event: ultimately, it was the fifth largest chat ever hosted by AOL. Eight thousand people logged on through AOL; another 10,000 tried to log on through other portals. Thirteen thousand questions were fired at us. Several thousand emails of support flooded our server. Television coverage reached 48 million people. There were Koko jokes on late night talk shows. Suddenly, we had a measure of the potential Koko might have as an "ambassador" for gorillas worldwide. She captures the public's imagination. The *New York Times* looked at our web site and did a front page story on her art. We have since been asked to do several exhibits. Public Television's Fred Rogers, of "Mr. Rogers' Neighborhood," came to visit Koko and aired a moving episode about diversity. A documentary of our work aired on PBS' "NATURE" program in 1999. "People to People," a citizen ambassador program, invited us to lead an Interspecies Communication Delegation to Africa to contribute to ongoing discussions of conservation needs and possibilities. Thus, gorillas in captive settings can spearhead conservation efforts directed toward their wild counterparts (Patterson and Gordon, this volume).

9.   CONCLUSION

This is the time to capitalize on the public fascination with these "charismatic megafauna." With growing awareness of the bushmeat trade threat, and growing understanding of the personhood of the great apes, we have the opportunity to instigate a paradigm shift: a radical evolution in our consciousness regarding the relationship we have with our close kin, the apes, and with animals and nature in general. Koko could significantly leverage this shift by engendering the emotional connection that is needed to motivate people to action. A teacher recently wrote to us: "After reading the picture book of Koko's life story, the children are hooked and feeling great compassion and love for her. When we think of saving the earth, we can't help but think of Koko." Could Koko become an icon of African rainforest protection, in the way Smokey the Bear has been for the prevention of forest fires in the United States? At the Gorilla Foundation, we are looking forward to helping to develop the multidisciplinary strategies that are required to save the great apes and their habitats, and to insure optimal living conditions for captive gorilla populations.

ACKNOWLEDGMENTS

We thank the organizers for bringing us together at the Great Apes of the World conference. We also thank Thomas Butynski, Anthony Rose, and Karl Ammann for sharing surveys and graphics with us. Ron Cohn, Kevin Connelly, Gary Stanley, and Wendy Gordon also contributed significantly to the development of this paper.

REFERENCES

Akers, J.S. and Schildkraut, D.S., 1985, Regurgitation/reingestion and coprophagy in captive gorillas, *Zoo Biology* 4: 99-109.

Boysen, S.T. and Butynski, T., this volume.

Dupain, J. and Van Elsacker, L., this volume.

Gould, E. and Bres, M., 1986, Regurgitation and reingestion in captive gorillas: Description and intervention, *Zoo Biology* 5: 241-250.

Harcourt, A., 1996, Is the gorilla a threatened species? How should we judge?, *Biol. Conservation* 75: 165-176.

IUCN, 2000, Red List of Threatened Species. Eletronic document, http://www.iucn.org, accessed January 2001.

Kenny, D.E., Cambre, R.C., Alvarado, T.P., Prowten, A.W., Allchurch, A.F., Marks, S.K., and Zuba, J.R., 1994, Aortic dissection: An important cardiovascular disease in captive gorillas (*Gorilla gorilla gorilla*), *J. of Zoo and Wildlife Medicine* 25(4): 561-568.

Meehan, T., 1997, Disease concerns in lowland gorillas. Pp. 153-159 in: (Eds. J. Ogden and D. Wharton), *Management of Gorillas in Captivity* The Gorilla Species Survival Plan and the Atlanta/Fulton Co. Zoo.

Morell, V., 1994, Will primate genetics split one gorilla into two?, *Science* 265: 1661.

Nishihara, T., 1996, Insect-eating by western lowland gorillas. Abstract #664, XVth Congress of the Inernational Primatological Society, Madison, Wisconsin.

Oates, J., 1996, *African Primates*, Gland: International Union for Conservation of Nature and Natural Resources.

Ogden, J. and Wharton, D. (Eds.), 1997, *Management of Gorillas in Captivity*, The Gorilla Species Survival Plan and the Atlanta/Fulton Co. Zoo.

Patterson, F.G.P. and Cohn, R.H., 1994, Self-recognition and self-awareness in lowland gorillas. Pp. 273-290 in: (Eds. S.T. Parker, R.W. Mitchell and M.L. Boccia), *Self-awareness in Animals and Humans*, New York: Cambridge University Press.

Patterson, F.G.P. and Gordon, W., this volume.

Rose, A.L., this volume.

Schulman, F.Y., Farb, A., Virmani, R., and Montali, R.J., 1995, Fibrosing cardiomyopathy in captive western lowland gorillas (*Gorilla gorilla gorilla*) in the United States: A retrospective study, *J. of Zoo and Wildlife Medicine* 26(1): 43-51.

Singer, P. and Cavalieri, P., this volume.

Tanner, J., 1998, Gestural communication in a group of zoo-living lowland gorillas, Unpublished doctoral dissertation, University of St. Andrews, Scotland.

Yerkes, R.M., 1927, The mind of a gorilla, *Genetic Psychology Monographs*, II (1 & 2).

Yerkes, R.M., 1927, The mind of a gorilla: Part II, Mental development, *Genetic Psychology Monographs*, II (6).

Chapter 11

TWENTY-SEVEN YEARS OF PROJECT KOKO AND MICHAEL

F.G.P. Patterson and W. Gordon
The Gorilla Foundation/koko.org, P.O. Box 620530, Woodside, CA 94062

1.   INTRODUCTION

Imagine that you are a gorilla hunter. You earn your living and support your family through the illegal but lucrative killing of gorillas, chimpanzees, and other endangered species for the commercial bushmeat trade (Ammann, 1996; Rose, 1996a, 1996b). Gorillas are large animals, your customers are willing to pay premium prices for their meat, and your attitude is: "Why should I not shoot these animals? They're meat." "Why should I feel bad for a gorilla? He is just a stupid animal" (McCrae, 1997: 75). Then imagine you learn more about one individual gorilla.

Koko communicates using sign language, and has a vocabulary of over 1,500 words. She also understands spoken English, and often carries on "bilingual" conversations, responding in Sign to questions asked in English. She is learning the letters of the alphabet and can read some printed words, including her own name. She uses a computer. She has achieved scores between 85 and 95 on the Stanford-Binet Intelligence Test.

She demonstrates self-awareness by engaging in self-directed behaviors in front of a mirror, such as making faces or examining her teeth, and by her appropriate use of self-descriptive language (Figure 34). She lies to avoid the consequences of her own misbehavior, and anticipates others' responses to her actions. She engages in imaginary play, both alone and with others. She has produced representational paintings and drawings. She remembers and can talk about past events in her life. She understands and has appropriately used time-related words such as, "before", "after", "later", and "yesterday".

Figure 34. Koko explores her mirror image.
© Dr. Rondald H. Cohn/Gorilla Foundation/Koko.org

She laughs at her own jokes and those of others. She cries when hurt or left alone, and screams when frightened or angered. She talks about her feelings, using words like "happy, sad, afraid, enjoy, eager, frustrate, mad, shame," and most frequently, "love." She grieves for those she has lost—a favorite cat who has died, a friend who has gone away. She can talk about what happens when one dies, but she becomes fidgety and uncomfortable when asked to discuss her own death or the death of her companions. She displays gentleness with kittens and other small animals. She has even expressed empathy for others seen only in pictures.

You then learn that this individual is Koko, a female western lowland gorilla (*Gorilla gorilla gorilla*), the same species as the gorillas you have been hunting. Does your new knowledge of her change your attitude about hunting gorillas? In at least one case so far, an introduction to Koko through books and photos has already had a profound effect on one former gorilla hunter (Rose, 1999).

2. OVERVIEW OF PROJECT KOKO AND MICHAEL

For nearly 30 years, Project Koko and Michael has been helping to answer questions about language, cognition, self awareness, and the biological bases of behavior in great apes and humans. Project Koko began in July 1972, when Koko was one year old. The original questions were: can a gorilla master the basics of symbolic communication at least as well as chimpanzees, and what might be the limits of this newly discovered symbolic ability in apes?

2.1   Background

R. Allen and Beatrice Gardner conducted the pioneering work with sign language communication with chimpanzee Washoe, who acquired 132 signs of American Sign Language (ASL) in 51 months (Gardner and Gardner, 1969, 1978). Initially Project Koko was structured quite similarly to Project Washoe—we sought to investigate many of the same language parameters: vocabulary development, generalization, semantic relations, comprehension, and productivity. Our aim was to create a body of data from which direct comparisons could be made between Koko and Washoe and Koko and human children (Patterson, 1980).

2.2   Results and Comparisons

2.2.1   Vocabulary

From the age of one year, Koko has been living and learning in a language environment that includes ASL and spoken English. Koko combines her working vocabulary of over 500 signs into statements averaging three to six signs in length. Her emitted vocabulary—those signs she has used correctly on one or more occasions—is over 1,000. Our analysis of Koko's linguistic development after the first ten years of the project indicated an emitted vocabulary of 876 signs (Patterson and Cohn, 1990). Koko was adding an average of 80 words each year between 1976 and 1979, and about 35 new words a year between 1980 and 1982. She has continued to both learn and invent new signs in the 19 years since.

2.2.2   Bilingualism

Although Koko was never explicitly instructed in English, no attempt was made to shield her from it, and most of her human companions speak English while signing. Koko's receptive vocabulary in English is estimated at several times her productive sign vocabulary. In formal tests of her language comprehension she responded equally well to questions asked in spoken English only or ASL only (Patterson and Linden, 1981).

2.2.3   Novelty

Koko has generated numerous novel signs without instruction, modulated standard signs in ASL to convey grammatical and semantic changes, used signs simultaneously, created compound names (some of which may be intentional metaphors), engaged in self-directed and noninstrumental signing, and has used language to refer to things removed in time and space, to deceive, insult, argue, threaten, and express her feelings, thoughts, and desires. These findings, together with documentation of her sign language acquisition and production capabilities, and her sign language and spoken English comprehension capabilities, support the conclusion that language acquisition and use by gorillas develops in a manner similar to that of human children, but at a slower rate.

2.2.4   Michael[1]

Koko is not alone in her linguistic accomplishments. Her multi-species "family" included Michael, a 26-year-old male gorilla. Although he was not introduced to sign language until the age of three and a half years, he learned to use over 400 different signs (Figure 35). Both gorillas initiate the majority of their conversations with humans and combine their vocabularies in creative and original sign utterances to describe their environment, feelings, desires, and even what may be their past histories. They also sign to themselves and to each other, using sign language to supplement their natural communicative gestures and vocalizations.

Sign language has become such an integral part of their daily lives that Koko and Michael are more familiar with the language than are some of their human companions. Both gorillas have been known to sign slowly and to repeat signs when conversing with a human who has limited signing skills. They also attempt to teach others as they have been taught. For example, one day Michael had been repeatedly signing "chase" (hitting two fisted hands together) but was getting no response from his companion, who did not know this sign. He finally took her hands and hit them together and then gave her a push to get her moving. Similarly, Koko has often been observed molding the hands of her dolls into signs. Outdoors, she once held a red blanket up before Ndume, her 19-year-old male gorilla companion, and signed "that red."

---

[1] Sadly, Michael died in April, 2000.

Figure 35. Michael signing "fruit."
© Dr. Ronald H. Cohn/Gorilla Foundation/Koko.org

2.2.5   Comparisons with Human Children

John Bonvillian and one of us (FGPP) compared Koko's early language development to that of 22 young human children of deaf parents learning sign as their first language. Although the human children learned new signs more rapidly than the gorillas, and fewer of their first-learned signs were iconic, there were many similarities between the two species in early sign vocabulary development. In comparisons of the first ten and first 50 vocabulary items acquired by the subjects, the content of Koko's and Michael's lexicons was quite similar to that of the children learning to sign, particularly at the 50-item vocabulary level (Bonvillian and Patterson, 1993). By the age of about 20 months, Koko began using her signs to label new instances of previously acquired concepts, indicating that the meanings of her signs were now based more on the identifying characteristics of the concepts than on specific concrete instances. The young children's initial steps in acquisition of meaning resembled Koko's. Koko's cognitive and language development during the first several years, although slower-paced, resembled that of the sign-learning children (Bonvillian and Patterson, 1997).

Like conversations with young children, in many cases conversations with gorillas need interpretation based on context and past use of the signs in question. Alternative interpretations of gorilla utterances are often possible. Even if the gorillas' use of signs does not meet a particular definition of language, studying their usage can give us a unique perspective from which to understand more directly their physical and psychological requirements. By agreeing on a common vocabulary of signs, we establish two-way communication between humans and gorillas. We can learn as much from what they say as we can by evaluating how they say it.

2.2.6   Self-Recognition

In 1990, Koko successfully passed a mirror self-recognition test, the first gorilla on record to clearly demonstrate this ability (Patterson and Cohn, 1994). Chimpanzees (*Pan*) and orangutans (*Pongo*) had already shown that they could recognize their mirror images, but the few gorillas previously tested had not, leading researchers to conclude that gorillas were somehow different from the other apes in this area of cognitive development (Gallup, 1987; Ledbetter and Basen, 1982; Suarez and Gallup, 1981). From the age of about four years Koko's self-directed behavior with mirrors indicated a capacity for self-recognition, but we felt it was also important to provide data strictly comparable to self-recognition studies done with other great apes.

Mirror self-recognition is generally considered to be one indicator of self-awareness, a capacity once assumed to be present only in humans. Koko has shown other abilities related to self-awareness, including the acquisition and use of personal pronouns and proper names, reference to her own internal and emotional states, attribution of mental states to others, self-conscious behaviors, value judgments, self-talk, humor, symbolic play, and expressions of intentionality, deception, and embarrassment (Figure 36).

3.   LOOKING FORWARD

There are many other aspects of cognition and language yet to be explored with Koko. Her unique situation and communication abilities make her an ideal subject for such studies. We are currently exploring her use of pretense (Matevia, *et al.*, in press), as part of the larger question of the gorilla's capacity for a theory of mind (see Singer and Cavalieri, this volume). Will the knowledge that gorillas are intelligent, self-aware, and capable—not only of experiencing thoughts and emotions but also of expressing them through language—be enough to prevent our species from obliterating them from the planet in the next few decades?

Figure 36. Koko with camera, next to Ron taking photo in a mirror.
© Dr. Ronald H. Cohn/Gorilla Foundation/Koko.org

Originally we expected to work with Koko for only about four or five years. As Koko's achievements with language exceeded all expectations, and as we became aware of her sensitivities and vulnerabilities as an individual, we realized that our commitment to Koko and the project—and to all gorillas—would be for life. Today, as we continue our study of interspecies communication and primate language ability with Koko, we find that our focus increasingly shifts toward conservation.

3.1   Koko as an Ambassador for Gorillas Worldwide

Over the years, it has become obvious that Koko can be a very effective ambassador for her species. Our book *Koko's Kitten* (Patterson, 1985), a children's book about Koko's gentleness with the kitten she named "All Ball" and her grieving over the kitten's death, has generated moving responses from the public ever since it was published. This simple story, with accompanying photos and videos, has proven to be a powerful means of affecting people's perception of gorillas. Thousands of letters and e-mails from children, teachers, and adults bear strong testament to the vital role that Koko plays as spokesperson and educator.

3.1.1   Koko's Kitten

Similarly, we endeavor to get our message out to those who now hold the fate of gorillas in their hands. Recently we translated *Koko's Kitten* into French, a language used widely in the African countries where free-living lowland gorillas still exist. Our plan is to develop an educational curriculum to go along with the books, and to distribute them to hunters, government officials, school children, and the city dwellers who consume commercial bushmeat.

We have already sent 10,000 copies of the book in French to Cameroon, and when it was shown to the gorilla hunter I described above, he was "visibly moved" by what he learned about the animals he had been killing for the bushmeat trade (Rose, 1998a). He looked up from the book several times to exclaim in an excited voice, "Is this really so?" (Rose, 1999). He expressed eagerness to show the book to his hunter friends and took several copies to share with them. Now, with support from the Bushmeat Project, this man is no longer hunting gorillas and has turned to conservation. Not only did he cease hunting apes, on at least two occasions he convinced another hunter not to shoot the gorillas they came across, even though it meant forfeiting the $80 they could have made by killing two silverbacks (Rose, 1998b).

3.1.2   Gorilla Art

Koko and Michael have produced some beautiful paintings over the years. Some of their paintings seem to be representational—they have painted from models and they have painted in response to requests to depict emotional states such as anger and love (Figure 37). We have written about their artwork in our journal, *Gorilla*, from time to time and have shown their paintings in a few exhibits—impressing a limited group of art lovers. However, when we recently added several of the gorillas' paintings to our Internet website we were able to reach a new segment of the public and to educate them about gorillas. People who may have had little interest in apes—their language capabilities, or their imminent demise in the wild— were suddenly drawn in by their artistic talents.

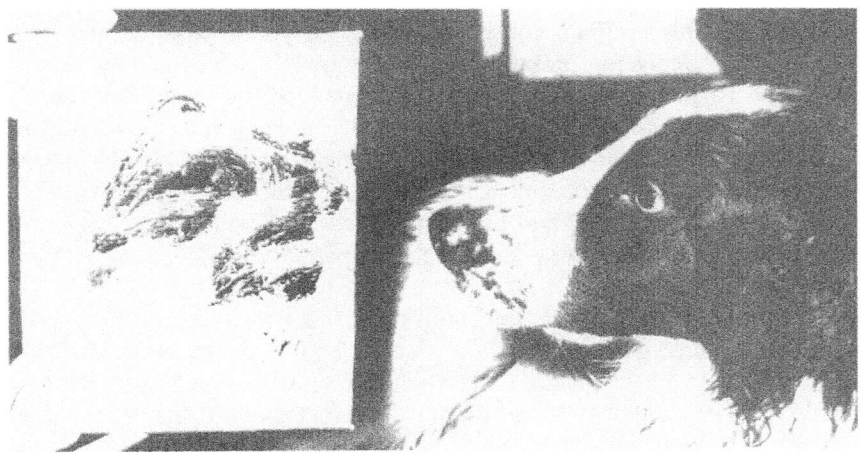

Figure 37. Michael's "Apple Chase" painting held next to the dog Apple for comparison. ©
Dr. Ronald H. Cohn/Gorilla Foundation/Koko.org

3.1.3   Broadcasting Koko's Image

We are also exploring additional ways we can utilize Koko's special appeal to help educate the world about gorillas and encourage their conservation. There are numerous things we can do that do not directly impact Koko or disrupt her daily life in any way. Our Internet website is one way that we share Koko's personality and accomplishments with the world.

In April 1998, Koko made an even more direct use of the Internet. The Internet service provider America Online hosted a unique chat session with Koko (the text of this chat is available on the website www.koko.org), which received a great deal of media attention both before and after the chat event, and was attended by tens of thousands of people from all over the world (see Patterson and Matevia, this volume). Ron Cohn videotaped while one of us (FGPP) asked Koko each question as they were relayed by phone. From Koko's perspective, the Internet chat session simply required her to spend about an hour of quality time with the two people who raised her. But the thousands of email messages we received in the next few days revealed just how wide an audience Koko was able to reach through this medium. There were messages from around the world, including Russia, Finland, Germany, England, Thailand, Japan, Australia, Holland, Portugal, Spain, Italy, Argentina, and Malaysia.

Koko reached out to another, different segment of the human population on 28 July 1998, when she was featured in an episode of "Mister Rogers' Neighborhood," a popular children's show on public television. The audience for this program consists primarily of young children and their parents and caregivers. Koko thoroughly enjoyed her videotaping session with Fred Rogers, host of the show. She spent well over an hour gently leading him around her room, hugging and grooming him, removing his trademark cardigan and his shoes, and demonstrating her ability to use the various props that he brought: a harmonica, a camera, a hat, and a beach towel with cats on it. When the suggested "script" called for a game of peek-a-boo with the towel, Koko readily participated, even for several retakes.

The theme of the episode was inclusion of those who are different. With Koko the featured guest on the show, this can be interpreted as the inclusion of gorillas (and by extension, other nonhumans) in the category of sentient beings whose individual lives matter. The images of Koko with Mister Rogers, like the images of her cradling tiny kittens, have had an impact on viewers young and old.

4.   CONCLUSION

Koko's unique relationship with humans and other animals elicits a strong emotional reaction and encourages people to identify with her. It is our hope that for at least some of those people, identification will be translated into action on behalf of gorillas and all the great apes.

As Tony Rose (1998b; this volume) points out, we humans are tribal primates when it comes to taking conservation action. It is the individuals we count as our close kin who come first with us. We must therefore do everything within our means to promote a very personal one-to-one connection with gorillas. To date, we have made use of the following to achieve this: stories, photos, and gorilla conversations in our semi-annual journal, *Gorilla*; personal replies to letters from the public; photos and biographies of our gorillas; examples of their artwork on our Internet web page; two AOL online chats with Koko; Koko's appearances on popular television shows; a unique TV public service spot in which Koko herself asked viewers to "help" gorillas; publications such as the book *Koko's Kitten* aimed at a general audience; and the ongoing free distribution of materials about gorillas for teachers and school children.

Our next projects include establishing a visitor center open to the public as an adjunct to our gorilla sanctuary on Maui, inclusion of a "Koko Cam" live video feed on our Internet website, licensing of various "Koko" products from dolls to software, and perhaps most importantly, increased free distribution of French-language editions of the books *Koko's Kitten* and *Koko's Story* throughout the African countries where gorillas still live. With

Koko as a charismatic ambassador helping to bridge the diminishing gap between human and gorilla, perhaps it will not be too late to stop the slaughter of the apes and ensure a future for these beings who truly are our kin.

Eternal Losses

The equatorial rainforests are the lungs of the planet.
Now filled with the smoke from man's fires,
they are burning away into oblivion.

In the foreground, a lowland gorilla cuts through our conscience
with a piercing gaze and asks—
why?
—*Schim Schimmel (1993)*

ACKNOWLEDGMENTS

Ron Cohn, Kevin Connelly, and Marilyn Matevia contributed significantly to the development of this paper. This paper is dedicated to the memory of both Michael our gorilla friend and Wendy Gordon. Michael inspired, amused, and awed us with his physical presence, beauty and genius. No words can express how affected we are by his untimely passing. Wendy was a treasured long-time friend, caregiver, and advocate of gorillas, who passed away 25 November 1998. She is deeply missed.

REFERENCES

Ammann, K., 1996, Primates in peril, *Outdoor Photographer* (February).
Bonvillian, J.D. and Patterson, F.G.P., 1993, Early language acquisition in children and gorillas: Vocabulary content and sign iconicity, *First Language* 13: 315-338.
Bonvillian, J.D. and Patterson, F.G.P., 1997, Sign language acquisition and the development of meaning in a lowland gorilla. Pp. 181-219 in: (Eds. C. Mandell and A. McCabe), *The Problem of Meaning: Behavioral and Cognitive Perspectives*, Amsterdam: Elsevier.
Gallup, G.G. Jr., 1987, Self-awareness. Pp. 3-16 in: (Eds. G. Mitchell and J. Erwin), *Comparative Primate Biology, Vol. 2, Part B: Behavior, Cognition, and Motivation*, New York: Alan Liss, Inc.
Gardner, R.A. and Gardner, B.T., 1969, Teaching sign language to a chimpanzee, *Science* 165: 664-672.
Gardner, R.A. and Gardner, B.T., 1978, Comparative psychology and language acquisition. In: (Eds. K. Salzinger and F. Denmark), *Psychology: The State of the Art*, Annals of the New York Academy of Sciences 309: 37-76.

Ledbetter, D.H. and Basen, J.A., 1982, Failure to demonstrate self-recognition in gorillas, *Am. J. Primatol.* 2: 307-310.

Matevia, M.L., Patterson, F.G.P., and Hillix, W.A., In press, in (Ed. R. Mitchell), *Pretending and Imagination in Animals and Children.*

McRae, M., 1997, Road kill in Cameroon, *Natural History* 106(1): 36-75.

Patterson, F., 1985, *Koko's Kitten*, New York: Scholastic Books.

Patterson, F.G., 1980, Innovative uses of language by a gorilla: A case study. Pp. 497-561 in: (Ed. K.E. Nelson), *Children's Language, Vol. 2*, New York: Gardner Press.

Patterson, F.G.P. and Cohn, R.H., 1990, Language acquisition by a lowland gorilla: Koko's first ten years of vocabulary development, *Word* 41(2): 97-143.

Patterson, F.G.P. and Cohn, R.H., 1994, Self-recognition and self-awareness in lowland gorillas. Pp. 273-290 in: (Eds. S.T. Parker, R.W. Mitchell, and M.L. Boccia), *Self-awareness in Animals and Humans*, New York: Cambridge University Press.

Patterson, F. and Linden, E., 1981, *The Education of Koko*, New York: Holt, Rinehart, and Winston.

Patterson, F.G.P. and Matevia, M.L., this volume.

Rose, A.L., 1996a, The African forest bushmeat crisis, *African Primates* 2(1): 32-34.

Rose, A.L., 1996b, The African great ape bushmeat crisis, *Pan African News* 3(2): 1-6.

Rose, A.L., 1998a, Conservation becomes a global social movement in the era of bushmeat primate kinship, *African Primates* IUCN/SSC, Atlanta.

Rose, A.L., 1998b, Finding paradise in a hunting camp: Turning poachers to protectors, *J. of the Southwestern Anthro. Assoc.* 38(3): 4-11.

Rose, A.L., 1999, On the road with a gorilla hunter: Turning poachers to protectors, *Gorilla Journal*, 21: 2-6.

Rose, A.L., this volume.

Schimmel, S., 1993, *Our Home, Too*, Canoga Park, CA: Collectors Editions, Inc.

Singer, P. and Cavalieri, P., this volume.

Suarez, S.D. and Gallup, G.G. Jr., 1981, Self-recognition in chimpanzees and orangutans, but not gorillas, *J. Hu. Evol.* 10: 175-188.

Chapter 12

WHO'S IN CHARGE? OBSERVATIONS OF SOCIAL BEHAVIOR IN A CAPTIVE GROUP OF WESTERN LOWLAND GORILLAS

T.A. Stein
Department of Anthropology, The University of Chicago, Chicago, IL 60637

1. INTRODUCTION

Gorillas (*Gorilla*) are unique among great apes in that they live in relatively stable bisexual groups containing more adult females than males. Solitary animals—usually fully adult males—make up about 10% of most populations (Stewart and Harcourt, 1987). The degree of cohesiveness of bisexual gorilla groups is also unique among apes (Fossey, 1972). In chimpanzee (*Pan*) and orangutan (*Pongo*) populations, the adults of each sex range separately; adult females and males do not live together in permanent social groups (Nishida and Hiraiwa-Hasegawa, 1987; Rodman and Mitani, 1987).

Schaller (1963) first stressed the cohesiveness of polygamous gorilla groups. He described the silverback as the hub of the group with females and youngsters clustered about him while other mature males remained peripheral to the group (Tuttle, 1986). "All group members appeared to be attuned to the silverback's activities; he leads them through their daily rounds of feeding, travel, resting and nesting. When extra silverbacks are with a group, its members usually react only to the dominant silver back as their leader" (Schaller, 1963: 238). Harcourt (1979) confirmed that the timing of group activities, the direction of travel, and the overall cohesiveness of the group is determined by the dominant adult male. He attributed the cohesiveness of the group to the relationships between each

177

adult female and the dominant silverback (Harcourt, 1979). Tuttle (1986: 295) noted that "the proximity of individuals, particularly adult females, to one another is commonly attributable to their attraction to the silverback instead of mutuality among themselves." In fact, the proximity of females to the dominant adult male is inversely proportional to the age of their offspring (Harcourt, 1979). In addition, immature gorillas are attracted to the dominant male. By age three years, immatures regularly seek proximity to the silverback and often leave their mothers to do so (Harcourt, 1978; Stewart and Harcourt, 1987). Overall, while the dominant male appears to be solely responsible for maintenance of daily operations and the cohesiveness of the group, the adult females and immatures are responsible for maintaining the relationship—through proximity—with the silverback.

During spring and summer 1994 the Brookfield group of lowland gorillas (*G. gorilla gorilla*) consisted of nine individuals with one silverback, an arthritic 38-year-old. Because he was the only silverback (the next oldest male was a nine-year-old blackback), it was assumed that he was the dominant male. However, although the blackback was young, he seemed to have a role, or was attempting to gain a role, in the leadership of the group. The 9-year-old male was much younger than any dominant male observed in the wild. Males are not considered fully adult until they are at least 15 years old (Stewart and Harcourt, 1987). In contrast, the 38-year-old silverback was older than dominant males documented in the wild.

I attempted to determine which male was dominant in the Brookfield group from the perspective of the gorilla group members. Assuming that the adult females and the immatures maintain a close social relationship with the dominant male, I analyzed the relationships of the members of the group with each of the two males to determine which was in fact responsible for cohesiveness and maintenance of the group. If the relationships between the members of the group and the dominant male constitute the cohesive factor in the overall social structure, then one should be able to determine which male is the dominant male *vis-à-vis* the members of the gorilla group by examining those relationships.

2. MATERIALS AND METHODS

2.1 Subjects and Setting

I collected data during a twelve-week period while observing the gorilla group at the Brookfield Zoo in Brookfield, Illinois. I collected 8.5 hours of data from the group in the zoo habitat area and 21 hours of data from the group in the holding area. Construction in the vicinity of the habitat area dictated that the group be kept in the holding area 24 hours per day from 25 April 1994 throughout the study period. The zoo habitat—or exhibit—is a quasinaturalistic indoor habitat with both natural (skylights) and artificial lighting (Figure 38). The holding area is a group of connected cages inside with completely artificial lighting; the gorillas can move freely from one cage to another (Figure 39).

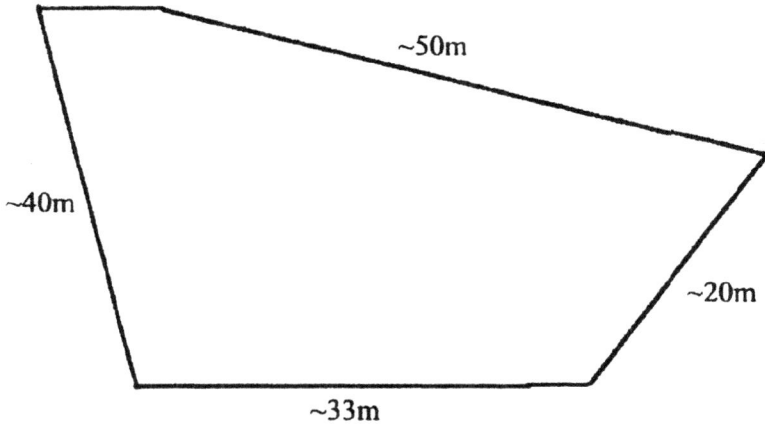

Figure 38. Approximate dimensions of gorilla exhibit area at Brookfield Zoo

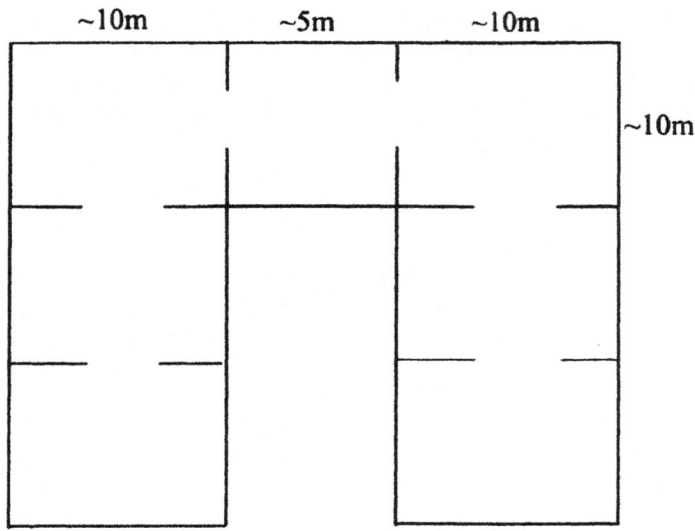

Figure 39. Approximate dimensions of gorilla holding area at Brookfield Zoo

The gorilla group consisted of nine individuals: one 38-year-old-silverback, one 9-year-old blackback, three adult females (ages 33, 20, and 12 years), one juvenile/subadult male (6 years), one juvenile/subadult female (6 years), one juvenile female (3.5 years), and one juvenile male (2.75 years). Six of the nine individuals are related to each other, but neither the silverback nor the blackback are related to other group members.

2.2   Procedures and Sampling Methods

I collected data using focal-animal and scan sampling methods (Altmann, 1974). All occurrences of the actions or interactions specified in the ethogram (Table 18) were recorded during 30-minute intervals for each of the two male subjects. Most data are from mornings between 09.30 and 12.00 hours. The time in which the 30-minute intervals occurred varied for each of the focal animals on a daily basis. I attempted to record any social behavior involving the focal animal, including which individual(s) initiated and terminated bouts of behavior and the reaction of the other individual(s). I recorded actions such as display behavior, proximity, play, touching, and agonistic behavior. The duration of the behavior was recorded whenever possible. If a behavior had not been terminated by the end of the 30-minute sampling interval, I recorded the duration thus far with a notation that it had not been terminated. If a behavior continued into and was terminated during a scan sample, I recorded the duration as >X and <Y, with the interval between X and Y equal to two minutes (the duration of the scan sample).

Table 18. Ethogram of Brookfield Lowland Gorilla Group

(Tch) touching:	any physical contact between two or more members of the social group that is not one of the other physical social behaviors
(Gr) grooming:	grooming behavior between two or more members of the group
(Pl) playing:	chasing and/or non-agonistic physical contact between two or more members of the group; and/or object manipulation, clapping, and/or rolling around by oneself
(AB) agonistic:	physical contact between two or more members of the group resulting from attack and involving hits, kicks, bites, or pinches; and/or threatening behavior in the form of pig-grunts, roars, growls, or screams directed toward a specific member or members of the group
(Dp) display:	behavior involving chest beating, throwing items, running, swinging on ropes/vines/cage, vocalizing (hooting), and/or ground beating
(Px) proximity:	two or more individuals within 3 m of each other while on exhibit or within 1 m of each other while in holding
(NA) no action:	individual with whom social behavior is being initiated or terminated does not react
(MA) move away:	individual with whom social behavior is being initiated moves away from initiator
(Fl) follow:	individual with whom social behavior is being terminated moves in the direction of the terminator within 2 to 3 seconds
(Rc) reciprocate:	individual with whom social behavior is being initiated or terminated reacts in the same way as the initiator/terminator
(Rs) resting:	not feeding, moving from one place to another, or playing
(Fd) feeding:	putting food and/or water into mouth
(Sk) suckling:	infant suckling on breast
(Lc) locomoting:	moving from one place to another that does not require climbing
(Cl) climbing:	moving up or down an obstacle (e.g., side of "hill", trees, cage, rope, etc.) in the habitat or holding area
(Bp) bipedal:	standing upright on two legs allowing at least one arm to be free (i.e., at least one hand is free while the other may be grasping side of hill, cage, other individual)
(St) sitting:	sitting where major source of support is provided by gluteal region
(Sq) squatting:	squatting where major source of support is provided by foot, including heel
(Qd) quadrupedal:	using both feet and hands (i.e., knuckles) for support
(Ly) lying:	more than hands, feet, and gluteal region in contact with surface for support

During each 30-minute interval, I interrupted observations every 10 minutes to execute a scan sample (Altmann, 1974) of each member of the group. The scan took ≤2 minutes. I recorded individual activities, neighbor(s) within a certain distance, and postural position of each individual during each scan (Table 18). When I recorded "lying" as a subject's postural position, I noted whether the individual was on its side, prone, or supine. In addition, I noted each individual's neighbor(s). On exhibit, a neighbor was considered to be ≤3 m, while in holding a neighbor was ≤1 m.

2.3   Analysis

I conducted separate analyses for each of the two focal animals in order to compare their social behaviors and interactions. I analyzed scan sampling separately from focal sampling data. I used Chi-square tests to determine the statistical significance of the differences between the focal animals' behaviors and of the differences in the behaviors of the other members of the group.

I used focal-sampling data to determine the overall frequency of social interactions between group members and the focal animal. The frequency was recorded as the number of interactions per 30-minute interval. Social interactions were grouped into agonistic, affiliative (e.g., grooming, playing, touching), neutral (e.g., proximity), and display behaviors.

I used records of which gorilla approached and which left during proximity behaviors to determine the frequency of displacements (i.e., animal initiates proximity and the receiver moves away) performed by and performed on each of the focal animals. These data were also used to determine who displaces whom and how often; i.e., the percent of the total displacements performed by a single individual. Frequency was recorded as the number of displacements per 30-minute interval.

I used focal-sampling data to determine responsibility for maintenance of proximity. The responsibility index (RI) for each group member with regard to the focal animal is defined as the percent of total proximity bouts that individual X initiates with the focal animal minus the percent of total proximity bouts that individual X terminates. If the index of individual X is positive, then s/he is more responsible for maintenance of proximity than is the partner (focal subject).

I used scan-sampling data to calculate the percent of time each individual was ≤3 m (on exhibit) or ≤1 m (in holding) of each of the focal animals. The number of scans that individual X was within the designated distance of the focal animal was divided by the total number of scans.

3. RESULTS

Data from the observation periods of the two focal animals (blackback as focal versus silverback as focal) were analyzed separately, as well as the data collected from scan samples. I treated the data from the exhibit together with the data from the holding area because there were too few observation periods in the exhibit to be treated separately.

3.1 Overall Social Interactions of the Blackback and the Silverback

The blackback was involved in many more social interactions than was the silverback. The average frequency of each category of behavior (e.g., proximity, display, affiliative, and agonistic) is greater for the blackback than the silverback (Figure 40). On average, the blackback was involved in proximity bouts and display behaviors more than affiliative or agonistic behaviors. The silverback, on the other hand, was only involved in proximity bouts and agonistic behaviors.

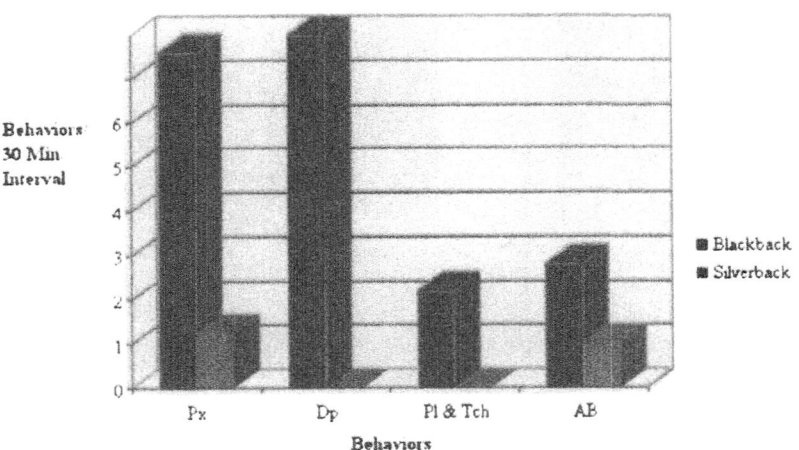

Figure 40. Average frequency of social interactions per 30-minute focal interval for each adult male. The Duckworth Test was used to determine statistical significance of the differences in frequencies between the blackback and the silverback for each of the behaviors. N=59 focal intervals: Px (L+R>7); Pl & Tch (L+R<7); AB (L+R<7).

3.2   Time in Proximity

Scan sampling data were used to determine the percent of time each member of the group spent in proximity to the blackback versus the time in proximity to the silverback.   Throughout all observation periods, the blackback and the silverback were never in proximity to each other. Therefore, there was no confusion determining with which male the other members of the group were involved during proximity bouts. I recorded a total of 175 scan samples. Figure 41 shows that both the adult females and the immatures spent more time in proximity to the blackback than to the silverback.

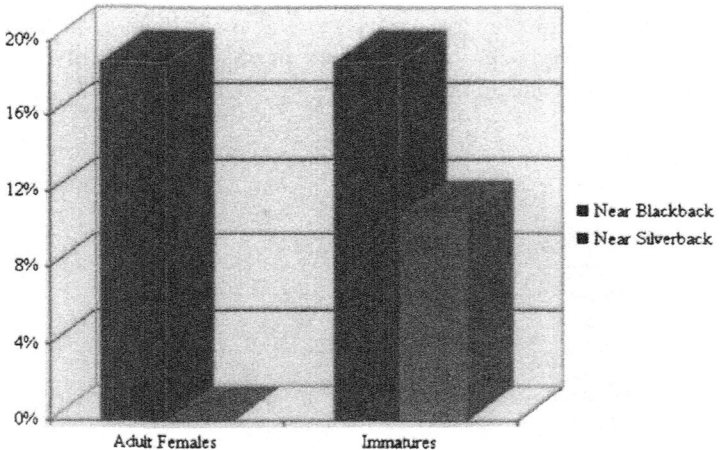

Figure 41. Percent of scan-sampling records (N=175) in which individuals were in proximity to blackback versus silverback. Chi-square tests were used to determine statistical significance of distribution for each group. Adult females: N=33; significant at p<0.001. Immatures: N=52; not statistically significant; p=0.052.

3.3   Responsibility for Proximity

Both the adult females and the immatures spent more time in proximity to the blackback than to the silverback. Figures 40 and 41 summarize focal sampling data of proximity bouts for the blackback and the silverback, respectively. The data are consistent with the scan sampling data in that the blackback was involved in many more bouts of proximity with the group members than was the silverback.

The blackback was involved in a total of 218 bouts of proximity while the focus of observation. The vast majority of the time spent in proximity to other members was with the adult females and the immature males (Figure 42). The relative percent of time that each member or group of individuals (e.g., adult females) initiated the bouts of proximity with the blackback is also indicated on the figure. The adult females initiated 84% of the bouts, the subadult male initiated 77% of the bouts, and the juvenile male and female initiated 100% of the bouts. In addition, the responsibility indices (RI) for the group members relative to the blackback are indicated on Figure 42. In all cases, the group member was responsible for maintaining proximity with the blackback.

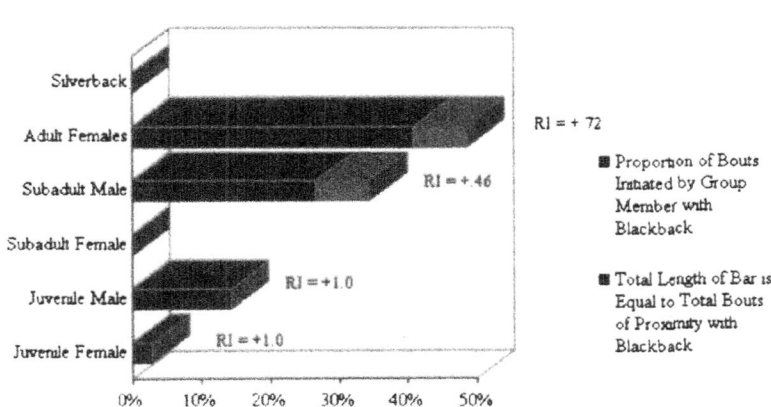

Figure 42. Distribution of bouts of proximity involving the blackback during focal-animal sampling. Chi-square tests were used to determine statistical significance of distribution. N=218 bouts of proximity; significant at p<0.001. Responsibility indices (RI) are recorded for each individual or group of individuals with regard to the blackback.

The silverback was involved in a total of 21 bouts of proximity over the entire observation period. Of those, the majority were spent in proximity to the subadult female (Figure 43). There was one instance each of a proximity bout with an adult female and with the subadult male. In all cases the group member initiated the proximity bout with the silverback; however, the group member also terminated the bout. The resulting RIs for all of the group members are zero, indicating that neither the group member nor the silverback was responsible for maintaining proximity.

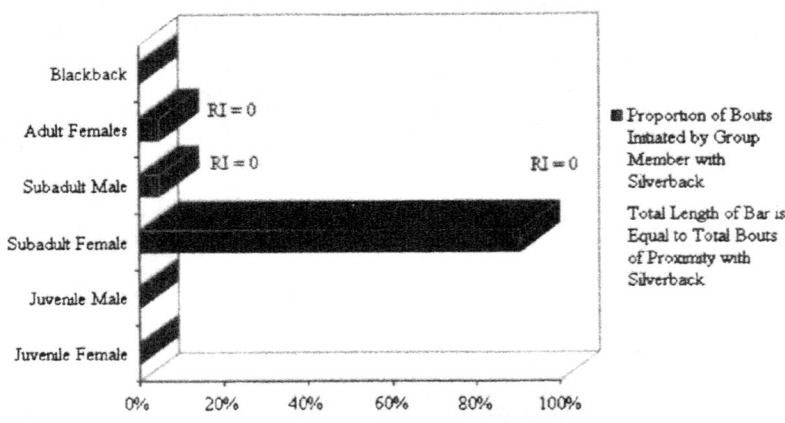

Figure 43. Distribution of bouts of proximity involving the silverback during focal-animal sampling. Chi-square tests were used to determine statistical significance of distribution. N=21 bouts of proximity; significant at p<0.001. Responsibility indices (RI) are recorded for each individual or group of individuals with regard to the silverback.

3.4   Approach-Retreat Interactions

I used focal-sampling data to determine individual reactions to non-aggressive approaches. If an individual approached another and the other moved away within one to two seconds, the "approached individual" was determined to have been displaced.

Figure 44 shows a distribution of all of the displacements in which the blackback was involved. In general, the blackback displaced other group members much more than he was displaced. He consistently displaced the immatures, and he displaced the adult females as often as they displaced him. However, the silverback displaced the blackback, while the blackback never displaced the silverback.

The silverback was never observed displacing or being displaced by a group member while he was the focus of sampling. The only displacements observed in which he was involved are those depicted in Figure 44.

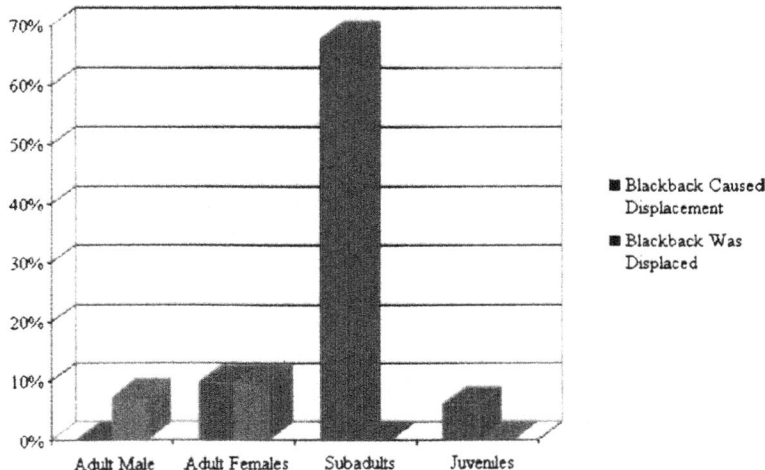

Figure 44. Distribution of displacements in which the blackback was involved. A Chi-square test was used to determine statistical significance of the difference between the total number of displacements caused by the blackback versus the total number of time the blackback was displaced. N=83; significant at p<0.001. It should be noted that the silverback was not involved in any displacements during observation while he was the focal animal.

4.   DISCUSSION

Ordinarily in naturalistic situations the silverback is the center of the bisexual gorilla group. He determines the timing of group activities, the direction of travel, and the overall cohesiveness of the group (Harcourt, 1979; Schaller, 1963). But what if the silverback is no longer capable of leading the group? Will a blackback take over if there is no other silverback to fill the role? How will the other group members react?

In this study, I observed social interactions, bouts of proximity, responsibility for proximity, and approach-retreat interactions between group members and the silverback and blackback males. The data indicate that the other members of the group respond to the blackback instead of the silverback as the leader of the group. The blackback was involved socially with the other members of the group much more than was the silverback (Figure 40). In particular, the blackback was involved in social interactions with all members of the group except the silverback, while the silverback was typically involved only with the subadult female.

4.1   Proximity Behaviors

According to Harcourt (1979), young males are of negligible importance in adult females' lives. The blackback at Brookfield is considered a young male by Harcourt's definition. However, the adult females interact with him and do not interact with the silverback. Specifically, the adult females focused their attention on the blackback via bouts of proximity with him (Figures 42 and 43). Harcourt (1979) argues that it is the relationships between the adult females and the leading male that maintain the cohesiveness of gorilla groups. Figure 41 demonstrates that the adult females in the Brookfield Zoo group were consistently in proximity to the blackback much more than to the silverback. In fact, an adult female was observed in proximity to the silverback only once (Figure 43). These data imply that the females chose the blackback over the silverback. Figure 42 also supports this assertion. Of the total number of bouts of proximity in which the adult females were involved with the blackback, the vast majority were initiated by the female. The RI of +0.72 indicates that most of the time the female did not terminate the proximity bout. Instead the blackback did, and often the female followed him. The adult females seem to be responsible for maintaining a relationship with the blackback. Again, according to the published literature, the adult females are responsible for proximity with the dominant male. By contrast, Figure 41 demonstrates that the Brookfield adult females ignored the silverback, and Figure 43 shows that the single time an adult female initiated a bout of proximity with the silverback, she also terminated it (RI=0).

The same situation seems to be true for the immature males. Harcourt and Stewart (1987; Harcourt, 1978) describe immature males as having an obvious and definite attraction to the dominant male. By age three years, immature males regularly seek proximity to the silverback, often leaving the proximity of their mother to do so. Although the data for immatures in Figure 41 are not statistically significant (p=0.052), 100% of the "near-silverback" column for immatures is attributable to a single individual. The subadult female tended to spend time in proximity to the silverback versus any other member of the group. On the other hand, the "near-blackback" column represents bouts of proximity of three of the four immatures. Figure

42 displays statistically significant data of the amount of time immature males spend in proximity to the blackback. Again the data indicate that immature males seek out the blackback and are responsible for maintaining proximity with him. Figure 43 illustrates the attraction of the single subadult female to the silverback. Only once was an immature male in proximity to the silverback. In addition, one should note that the RIs are zero in all cases. Like the adult females, immatures are not responsible for maintaining proximity to the silverback.

4.2   Approach-Retreat Interactions

Figure 44 depicts the percent of time that the blackback displaced each group member compared to the percent of time that the group member displaced him. Overall, the blackback displaced others more than he was displaced. However, while he displaced immatures consistently, he was displaced as often by the adult females as he displaced them. Moreover, in the only displacements (n=2) that involved the silverback, he displaced the blackback.

5.   CONCLUSIONS

The data concerning total social interactions of the Brookfield Zoo blackback compared to the silverback indicate that the blackback is more involved with the group members than is the silverback. Data of proximity bouts and responsibility for maintenance of proximity show that the blackback may in fact be the hub of the group instead of the silverback. The Brookfield data strongly suggest that the group members—adult females and immature males—react to the blackback as one would expect them to react to the leading male of the group. In addition, none of the data indicate that group members interact with the silverback as the leading male. After the data were collected (autumn 1994), handlers reported two separate instances of the 20-year-old female sexually soliciting the blackback. In 1995, she gave birth. Through genetic testing, the infant was identified as the offspring of the blackback. In summary, it appears as if the age distribution of the group, especially of the males, may have caused the blackback to assume responsibilities much earlier than he would have in the wild.

This study is of particular interest for the maintenance of captive groups of gorillas. In the absence of a silverback, these data suggest that a young blackback can successfully assume the leadership of a gorilla group.

ACKNOWLEDGMENTS

I would like to thank the Brookfield Zoo for allowing me to observe the gorillas. I am especially grateful to the Stephanie for all of her help and valuable suggestions during this project. I would also like to thank Dr. J. Altmann for advising me throughout this project.

REFERENCES

Altmann, J., 1974, Observational study of behavior: Sampling methods, *Behaviour* 49: 227-267.
Fossey, D., 1972, Living with mountain gorillas. Pp. 208-229 in: *The Marvels of Animal Behavior, National Geographic Society*, Washington, D.C.
Harcourt, A.H., 1978, Activity periods and patterns of social interaction: A neglected problem, *Behaviour* 66: 121-135.
Harcourt, A.H., 1979, Social relationships between adult male and female mountain gorillas in the wild, *Animal Behaviour* 27: 325-342.
Nishida, T. and Hiraiwa-Hasegawa, M., 1987, Chimpanzees and bonobos: Cooperative relationships among males. Pp. 165-177 in: (Eds. B.B. Smuts, D.L. Cheney, R.M. Seyfarth, R.W. Wrangham, and T.T. Struhsaker), *Primate Societies* Chicago: University of Chicago Press.
Rodman, P.S. and Mitani, J.C., 1987, Orangutans: Sexual dimorphism in a solitary species. Pp. 146-154 in: *Primate Societies*. (Eds. B.B. Smuts, D.L. Cheney, R.M. Seyfarth, R.W. Wrangham, and T.T. Struhsaker), *Primate Societies* Chicago: University of Chicago Press.
Schaller, G.B., 1963, *The Mountain Gorilla*, Chicago: University of Chicago Press.
Stewart, K. J. and Harcourt, A.H., 1987, Gorillas: Variation in female relationships. Pp. 155-164 in: *Primate Societies*. (Eds. B.B. Smuts, D.L. Cheney, R.M. Seyfarth, R.W. Wrangham, and T.T. Struhsaker), *Primate Societies* Chicago: University of Chicago Press.
Tuttle, R.H., 1986, *Apes of the World*, New Jersey: Noyes.

SECTION FIVE

PHYSIOLOGICAL BASES FOR BEHAVIOR AND AGING: GREAT APES AND HUMANS

INTRODUCTION

At least since the times of Thomas Huxley, scientists have recognized the close biological similarities of humans and African apes (Zihlman, 1997). From the apes' perspective, one negative consequence is their use as biomedical models of human diseases and conditions. However, the lines between human and nonhuman have blurred during the last three decades, and the public has increasingly questioned the morality of the biomedical use of apes and other species (Cavalieri and Singer, 1993; Singer and Cavalieri, this volume).

During the 1970s, considerable debate centered on the ethics of maintaining captive collections of animals and concerns about how exhibit animals were acquired. The zoo community attempted to respond to these issues, and conservation became an increasing focus of many of the world's best zoos. This shift in emphasis was accompanied by the realization that captive collections should be managed cooperatively across all institutions that hold that species. By 1980, this sensibility had coalesced so that American zoos claimed to "...focus efforts of many different institutions into a single, coherent program for conservation through research, publication, education, reintroduction, and fieldwork" (American Zoo and Aquarium Association, 2001: 4). Species Survival Plans (SSPs) were formed to help coordinate the efforts of various institutions holding a particular species (for example, to provide recommendations about breeding) while assisting with the conservation of that taxon in the wild. As public opinion opposing experimentation on apes gathered momentum, zoos increasingly became centers of noninvasive research where questions relevant to primate health and behavior could be explored and coordinated through SSPs. Such research could be carried out while fostering the conservation and/or well-being of the study subjects.

The chapters of section five reflect these recent trends in laboratory and zoo settings. Section five begins with three chapters describing preliminary results from studies coordinated through the Great Ape Aging Project. This multi-disciplinary project was created in response to increasing numbers of old great apes living in captivity in North America. For the first time in great ape evolutionary history, many captive individuals are surviving to the species' maximum life span—lives long enough to develop or express age-related diseases and conditions. However, little data exist on the aging process of apes or other primates (DeRousseau, 1994), on how to adjust captive maintenance as the ape ages, or on what kinds of behavioral and physical changes to expect as apes approach their species' maximum life span. The apes' close evolutionary relationship to humans allows addressing these gaps in our knowledge to enhance understanding of the human aging process.

Joseph Erwin and his colleagues provide an overview of the Great Ape Aging Project, which involves inter-institutional cooperation in data collection and the care given to the 45+ great apes in America that are 40 years old or older. They note that the non- or minimally-invasive techniques used to monitor the ape subjects will produce data that will benefit the aged individuals, will enhance our understanding of old wild apes, and will shed light on conditions usually associated with aging in humans. Robert Martens and his colleagues next describe spontaneously occurring diabetes in great apes, a condition associated with obesity and advancing age in both humans and great apes. While important for our understanding of diabetes in humans, these data can also improve the lives of captive diabetic apes. Daniel Perl and his colleagues describe preliminary results of their study of aging great apes' brains. Brains are collected after the natural death of captive apes, and subsequent analysis of neural tissue indicates that some of the pathological changes that occur in humans with Alzheimer's Disease are also present in the brains of aged great apes. This information illuminates aging processes in both humans and great apes and furthers our understanding of primate cognitive processes, particularly through comparative neuroanatomy.

Jurke and his colleagues continue this section with detailed analyses of ovarian hormones present in the feces and urine of female bonobos (*Pan paniscus*). Through the Bonobo SSP, noninvasive techniques are used to gather data from captive female bonobos housed at zoos in the US, which required collaboration and cooperation in data collection by personnel at all participating zoos. Jurke and his research team monitored females' hormone levels, which are interpreted in light of female sexual swelling changes and the sexual behavior of female and male bonobos in each enclosure. They find that urinary hormones can be used to monitor ovarian cycles. They also note that young male bonobos tend to copulate with females throughout the ovarian cycle, while older adult males may concentrate copulations at the time of peak estrogen. In an additional elaboration on section five's theme of ape aging, Jurke and his colleagues note that old female bonobos continue to cycle rather than undergoing menopause. This finding lends support to Pavelka and Fedigan's (1991) argument that human menopause is a unique, nonadaptive consequence of greatly increased life span in hominids.

Ronald Nadler closes section five with a consideration of the impact of age on male chimpanzee (*Pan troglodytes*) sexual arousability. His data support the hypothesis that sexual arousability declines with advancing male age. Younger males are more readily and frequently aroused, which may account in part for a young male's tendency to initiate mating throughout the ovarian cycle and with little regard to the female's sexual interest. Nadler notes, however, that the dominance interactions and familiarity of the mating pair are of great importance in the timing of copulation and may be key factors explaining individual deviations from more typical age and sex mating patterns.

REFERENCES

American Zoo and Aquarium Association, 2001, A brief history of the American Zoo and Aquarium Association. Electronic document, http://www.aza.org, accessed January 2001.

Cavalieri, P. and Singer, P. (editors), 1993, *The Great Ape Project: Equality Beyond Humanity*. New York: St. Martin's Press.

DeRousseau, C.J., 1994, Primate gerontology: An emerging discipline. Pp. 127-153 in: (Eds. D.E. Crews and R.M. Garruto), *Biological Anthropology and Aging*, Oxford: Oxford University Press.

Pavelka, M.S.M. and Fedigan, L.M., 1991, Menopause: A comparative life history perspective, *Yearbook of Physical Anthropology* 34: 13-38.

Singer, P., and Cavalieri, P., this volume.

Zihlman, A., 1997, African apes. Pp. 17-23, in (Ed. F. Spencer), *History of Physical Anthropology, Volume 1 (A-L)*, New York: Garland Publishing, Inc.

Chapter 13

THE GREAT APE AGING PROJECT: A RESOURCE FOR COMPARATIVE STUDY OF BEHAVIOR, COGNITION, HEALTH, AND NEUROBIOLOGY

J.M. Erwin[1], M. Bloomsmith[2], S.T. Boysen[3], P.R. Hof[4], R. Holloway[5], L. Lowenstine[6], R. McManamon[2], D.P. Perl[4,7], W. Young[8], and A. Zihlman[9]
[1]*Division of Neurobiology, Behavior, and Genetics, BIOQUAL, Inc., 9600 Medical Center Drive, Rockville, MD 20850; Foundation for Comparative and Conservation Biology*

[2]*Zoo Atlanta, Atlanta, GA*

[3]*Comparative Cognition Project, Ohio State University, OH; Living Links Center for Human and Ape Evolution and Behavior, Emory University, Atlanta, GA; Great Ape Aging Project, Diagnon Corporation, Rockville, MD*

[4]*Arthur M. Fishberg Research Center for Neurobiology, Mount Sinai School of Medicine, New York, NY*

[5]*Anthropology Department, Columbia University, New York, NY*

[6]*School of Veterinary Medicine, University of California, Davis, CA*

[7]*Neuropathology Division, Mount Sinai School of Medicine, New York, NY*

[8]*Scripps Research Institute, La Jolla, CA*

[9]*Anthropology Department, University of California, Santa Cruz, CA*

1.   INTRODUCTION

Great apes seldom survive in nature to reach old age, which has been described for chimpanzees as beginning by about 33 years of age (Goodall, 1986). Captive great apes often live much longer than is probably typical in the wild. There are likely many reasons for this, but not all are well understood. In captivity they are protected from predators, but death due to predation is uncommon in wild great apes. Diseases and infections are often treated effectively in captivity. Nutritional needs are typically met or exceeded in captivity, but this also leads to obesity and diabetes in some cases (see Martens, *et al.*, this volume). Captive apes are more likely than wild ones to receive fresh and uncontaminated water. Social incompatibilities are seldom allowed to lead to lethal combat in captivity. Consequently, some captive great apes live into their fourth, fifth, or sixth decade.

Living far beyond the life span typical of populations across tens of thousands of years of evolutionary adaptation is a recent phenomenon for humans, and is even more recent in the great apes. As in humans, longer life for great apes can result in increases in the incidence of age-related disorders. Increases in the *quantity* of life are not necessarily accompanied by increases in the *quality* of life. In fact, the genetically based structural and functional similarities between great apes and humans dictate shared risk of age-related disorders, including Type II diabetes, osteoporosis, cardiovascular disease, stroke, arthritis, benign prostatic hypertrophy, and possibly neurodegenerative disorders such as Alzheimer's disease (see Perl, *et al.*, this volume).

The fact that apes in human care live longer than their counterparts in the wild presents an expanded obligation to address quality-of-life issues. There is broad agreement that "convenience" euthanasia is inappropriate for great apes; consequently, there is no easy way of dealing with the disabilities that inevitably occur when life is extended. The obligation to provide high-quality care for the oldest apes carries with it an opportunity. The increased attention and care elderly apes may require can be supported in part by systematic but noninvasive monitoring of health and behavior. The information obtained can be applied to improved care of the individuals involved, to other members of their species, and, in some cases, to a more general understanding of aging in primates, including humans. The fact that the monitoring of health, behavior, and cognition among apes is fundable primarily for its relevance to human health does not diminish the benefit of this information to individual apes. The Great Ape Aging Project is designed to assure that we learn what we can about aging and age-related disorders without harm to the individuals studied—and with the potential for enhancement of their well-being and that of their species.

2.   METHODS

The Ape Taxon Advisory Group (APE-TAG) of the American Zoological Association (AZA) was contacted to seek information on the identity and location of the oldest great apes in American zoological gardens. Each of the Species Survival Plan (SSP) Species Coordinators provided lists of the oldest apes. Biomedical institutions holding great apes (primarily chimpanzees) were also contacted, including Primate Foundation of Arizona (Tempe); Coulston Foundation (Alamogordo); Southwest Foundation for Biomedical Research (San Antonio); University of Texas M. D. Anderson Cancer Center Science Park (Bastrop); University of Louisiana, Lafayette, New Iberia Research Center (New Iberia); and the Yerkes Regional Primate Research Center at Emory University (Atlanta). Retirement sanctuaries were also contacted, including Primarily Primates (Texas) and Wildlife Waystation (California). Communication research projects were surveyed to determine whether any of the apes in their charge were 40 years of age or older. The International Species Inventory System (ISIS) was contacted for an additional check on elderly great apes. All the zoos, research centers, and sanctuaries holding great apes 40 years of age or older agreed to participate in the initial phase of the Great Ape Aging Project. Following determination during Phase I that the project was feasible, our group began developing the project and its methods. Mollie Bloomsmith is developing the behavioral monitoring system. Sally Boysen and her students are developing the cognitive assessment system. Linda Lowenstine and Rita McManamon are developing a standard necropsy protocol for great apes in zoos. Among other things, they will use the human brain recovery protocol (for victims of Parkinson's disease) developed by Dan Perl as they develop the protocol for great apes. Warren Young is the project consultant for developing digital imaging capacities for the project. Ralph Holloway is advising regarding the creation of brain casts and possibly skull endocasts to go with them. Adrienne Zihlman is providing advice regarding skeletal aging and postural assessment from videographed locomotor behavior. Neuroanatomical studies of the brain specimens obtained by the project are in progress by Patrick Hof and his colleagues, and some of these results are described below. The methods used include precise histological preparation of thin sections and use of several kinds of staining, especially Nissl staining and some stains using calcium binding proteins. Quantitative stereological methods are also used to provide brain cell population counts and measures of neuronal volume. The methods involved in these multiple approaches are in the process of being developed. As these are developed, they will be described in publications reporting specific results.

3.   RESULTS

During 1997, 45 great apes over 40 years old were identified in American zoological gardens, research institutions, and sanctuaries. These included 29 chimpanzees, two bonobos, nine gorillas, and five orangutans. All the bonobos and orangutans, and most of the chimpanzees and gorillas, were females. This result suggests that survivorship in the great apes, as in humans, is greater in females than in males. As indicated above, the monitoring of behavior, cognition, and locomotion is still in the developmental phase, and health records have not been obtained for many of the great apes involved in the study. Participating institutions have provided 30 great ape brain specimens, some of which were already archived prior to the onset of the study. These have been prepared and studied in Patrick Hof's neurobiology of aging laboratory.

Some of the results of these neurobiological studies are included in Perl, *et al.* (this volume), and others have been reported elsewhere (Hof, *et al.*, 1998). Additional studies have also made use of these specimens (Gannon, *et al.*, 1998; Kheck, *et al.*, 1998). The most detailed published report is that of Nimchinsky, *et al.* (1999). Studies related to human neuropathology of aging have had the highest priority, such as assessments of neuronal loss in the entorhinal cortex as reported by Dr. Perl and colleagues (this volume) and in the anterior cingulate cortex. The anterior cingulate cortex includes large spindle cells that are diminished by about 60% in human victims of Alzheimer's disease (Nimchinsky, *et al.*, 1995). Examination of rhesus macaque (*Macaca mulatta*) brains revealed no spindle cells at any age, and it was widely believed that these cells were unique to humans. Examination of the anterior cingulate cortex of orangutans revealed that spindle cells were sparse but definitely present. They were more abundant in gorillas and even more common in chimpanzees. In bonobos, the abundance approached that found in humans, and a clustering pattern previously observed only in humans was also noted. This cell type was not found outside of the great ape group. The anterior cingulate cortex of 28 primate species were examined, along with more than 30 nonprimate mammals. A second unique type of cell, a medium-sized pyramidal cell, was found in the same area, and appeared also to be unique to humans and great apes.

4. DISCUSSION

The results of the comparative neurobiology studies are especially intriguing, not just because neurons were revealed that suggest that great apes could be vulnerable to Alzheimer's disease, but because the unique types of neurons detected in anterior cingulate cortex may confer on great apes some cognitive capacities not seen in animals other than humans and great apes (see also MacLeod, *et al.*, this volume). Stroke damage to human anterior cingulate cortex is sometimes associated with what might be called a "loss of personhood" or a "loss of visual self-awareness." The anterior cingulate cortex is implicated in many aspects of cognition (attention, memory), sensation (pain and speech perception), and motor function (vocal production and autonomic motor control). One special area of interest to primatologists is the prospect that anterior cingulate cortex is involved in mirror-image self-recognition or visual self-awareness. This type of self-awareness has been demonstrated in some great apes and in humans (see Gallup, 1987 for a review; Patterson and Gordon, this volume).

Much remains to be learned about the processes of normal aging and the neuropathology of aging in humans and other primates. We believe it is possible to use entirely noninvasive methods to study these processes and that the knowledge gained will promote the well-being of great apes, humans, and all other primates. In addition to the increased understanding of aging processes, it is clear that the project is already enabling new discoveries regarding fundamental neuroanatomy. The findings regarding neuronal types unique to great apes and humans, along with quantitative data on cell populations and cell volumes, suggest a role for modern comparative neurobiology in cladistic systematics, as well as an increased understanding of brain evolution.

ACKNOWLEDGMENTS

The authors gratefully acknowledge support for the project from USPHS/NIH grant AG14308 from the National Institute on Aging to BIOQUAL, Inc., J. Erwin, Principal Investigator, and to the Mount Sinai School of Medicine and all other participating organizations. We thank Randy Fulk (chimpanzee), Gay Reinartz (bonobo), Dan Wharton (gorilla), and Lori Perkins (orangutan) for providing lists of old apes.

REFERENCES

Gallup, G., 1987, Self-awareness. Pp. 3-16 in: (Eds. G. Mitchell, and J. Erwin), *Comparative Primate Biology, Vol. 2: Behavior, Cognition, and Motivation*, New York: Alan R. Liss.

Gannon, P., Broadfield, D., Kheck, N., Hof, P., Braun, A., and Erwin, J., 1998, Brain language area evolution, I: Anatomic expression of Heschl's gyrus and planum temporale asymmetry in great apes, lesser apes, and Old World monkeys. *Society for Neuroscience Abstracts* 24: 160.

Goodall, J., 1986, *The Chimpanzees of Gombe*, Cambridge: Harvard/Belknap.

Hof, P., Nimchinsky, E., Perl, D., and Erwin, J., 1998, Identification of neural types in cingulate cortex that are unique to humans and great apes, *Am. J. Primatol.* 45(2): 184-185.

Kheck, N., Gannon, P., Hof, P., Braun, A., and Erwin, J., 1998, Brain language area evolution, II: Human-like pattern of hemispheric asymmetry in planum temporale of chimpanzees, *Society for Neuroscience Abstracts* 24: 160.

Martens, R., Couch, R., Hansen, B., Howard, C., Kemnitz, J., Perl, D.P., Erwin, J.M., this volume.

MacLeod, C., Zilles, K., Schleicher, A., and Gibson, R., this volume.

Nimchinsky, E., Vogt, B., Morrison, J., and Hof, P., 1995, Spindle neurons of the human anterior cingulate cortex, *J. of Comp. Neurology* 355: 27-37.

Nimchinsky, E., Gilissen, E., Allman, J., Perl, D., Erwin, J., and Hof, P., 1999, A neuronal morphologic type unique to humans and great apes, *Proceedings of the National Academy of Science* 96: 5268-5273.

Patterson, F.P.G., Gordon, W., this volume.

Perl, D., Hof, P.R., Nimchinsky, E.A., Erwin, J.M., this volume.

Chapter 14

AN INTERNATIONAL DATABASE FOR THE STUDY OF DIABETES, OBESITY, AND AGING IN GREAT APES AND OTHER NONHUMAN PRIMATES

R. Martens[1], R. Couch[2], B. Hansen[3], C. Howard[4], J. Kemnitz[5], D.P. Perl[6], and J.M. Erwin[1]

[1]*Division of Neurobiology, Behavior, and Genetics, BIOQUAL, Inc., 9600 Medical Center Drive, Rockville, MD 20850*

[2]*Couslton Foundation, Alamogordo, NM*

[3]*University of Maryland School of Medicine, Baltimore, MD*

[4]*University of Northern Colorado Research Corporation, Greeley, CO*

[5]*University of Wisconsin Regional Primate Research Center, Madison, WI*

[6]*Neuropathology Division and Arthur M. Fishberg Research Center for Neurobiology, Mount Sinai School of Medicine, New York, NY*

1. INTRODUCTION

The project described here grew out of our personal experiences and conversations with zoo curators, primate keepers, primate veterinarians, and diabetes researchers regarding spontaneously occurring diabetes in great apes and other primates. It involves the convergence of several lines of inquiry concerning the health and well being of nonhuman primates and human public health. These include:

1. reports of clinical diabetes in great apes and concerns about managing the disease (chimpanzee, Rosenblum, et al., 1981; orangutan, Kemnitz, et al., 1994);

2. studies of spontaneous diabetes in macaques (e.g., Sulawesi crested macaque, Howard, 1972; and rhesus, Hansen, 1995) and collaborative field studies of populations and diabetes risk in nonhuman primates (e.g., Sulawesi macaques, Howard, et al., 1999; Sugardjito, et al., 1989).

The fundamental premise of this effort is that the causes and consequences of diabetes are identical or very similar in humans and other primates. All the available evidence supports this view. A fuller understanding of diabetes, along with improved methods of prevention, diagnosis, and management of this disease, is in the best interests of all who may be affected by it. A systematic study of clinical cases of diabetes (and pre-diabetes) in the great apes and other primates should prove beneficial for human public health, and it is also in the best interests of the focal nonhuman individuals and populations. All too often, diabetes is not diagnosed until it is so advanced that a clinical crisis occurs and some of the damaging consequences of diabetes are beyond reversal.

The complications of diabetes can be severe and debilitating and may involve damage to a variety of organ systems. Diabetes-associated disorders include cardiovascular disease, kidney disease, arthritis, neuropathy, and ocular complications such as retinopathy, glaucoma, and cataracts. The incidence of diabetes in nonhuman primates is not yet known, but it is one of the most common diseases in humans. It occurs in about 16 million Americans (about 6% of the population); however, only about half of the existing cases have been diagnosed. The others are progressing without diagnosis (Miller, 1998). Interestingly, two diagnostic studies of great apes in which glucose tolerance tests were administered found as many previously undiagnosed cases as had already been diagnosed (Kemnitz, et al., 1994; Rosenblum, et al., 1981). For example, Kemnitz, et al. (1994) found advanced diabetes in four out of nine orangutans tested, only two of which had been previously diagnosed.

Several kinds of diabetes exist in humans. The two most common are called Type 1 (juvenile or insulin-dependent diabetes) and Type 2 (adult-onset or noninsulin dependent diabetes). Both occur in the great apes, but, as in humans, most cases are Type 2. Despite the name "noninsulin dependent," Type 2 diabetes often progresses to a state in which insulin therapy is essential. Many cases are controlled through exercise and special diets—if they are diagnosed early. A chimpanzee at Taronga Zoo in Sydney, Australia, and an orangutan at Chicago's Brookfield Zoo were maintained on daily insulin injections for many years, but questions often arise regarding the best course of management. The study described here is an attempt to identify as many cases as possible and assist with management strategies.

2.   METHODS

2.1   Animals

The primary concern of the study is with the great apes, including chimpanzees (*Pan troglodytes*), bonobos (*P. paniscus*), gorillas (*Gorilla gorilla*), and orangutans (*Pongo pygmaeus*). Some additional effort will be made to include data from gibbons and siamangs (Hylobatidae) and Old World monkeys, especially rhesus macaques (*Macaca mulatta*), Sulawesi macaques (*M. nigra* and *M. maurus*), and patas monkeys (*Erythrocebus patas*). Some data sets already exist for the monkey species.

2.2   Data Management

Zoological gardens and research centers in which diabetic great apes and other primates reside (or resided prior to death) will be asked to contribute data from clinical records of affected and unaffected individual primates. The data will include, but will not be limited to, the following categories for each individual.

- Subject variables: species, subspecies (if appropriate and known), birth date (or estimated birth year), gender, father (if known), mother (if known), siblings (half or full), offspring, blood-type, rearing history, experimental history, clinical history (known illnesses and/or surgeries), and other relevant individual characteristics
- Clinical measures: weight, fasting glucose, insulin response, triglycerides, cholesterol, lipoproteins (HDL, MDL, LDL), albumin, and HDPE.

The data will be entered into a relational database program (not yet selected) on a personal computer with adequate memory to accommodate what can become a substantial and complex database. Data will be coded and securely stored in accordance with the confidentiality requirements of participating institutions.

Data analysis will include retrospective examination of the records of confirmed clinical cases of diabetes in great apes and other primates to identify the point at which values for each individual diverged from the normal range. Special attention will be given to the variables that are consistently diagnostic of diabetes. Efforts will also be made to identify the lag time from the earliest diagnostic clues to full onset of diabetes and to the onset of specific diabetes-associated disorders or disabilities. Combinations of potential indices will be assembled and their relative predictive power will be evaluated.

2.3   Postmortem Examination

Following natural death, a necropsy will be performed as soon as possible and major organs will be removed and preserved for detailed pathology assessments. Brain, eyes, aorta, pancreas, liver, kidneys, and other tissues potentially affected by diabetes will be obtained and included in a tissue bank that will key specimens to the detailed database of individual characteristics and clinical measures. The procedures used will be essentially the same as those undergone by human participants in tissue bank programs.

2.4   Surveys

Two initial surveys are planned. The first will be of zoological gardens and the second will be of research institutions. A brief questionnaire will request basic information regarding known or suspected cases of diabetes in great apes or other primates, the number of cases, and the methods of management. Information will also be sought on diabetes-associated disorders, whether or not a diagnosis of diabetes has been made. Each institution's willingness to allow a diagnostic glucose tolerance test will be ascertained.

3.   DISCUSSION

A proposal to fund the project described here has been prepared and submitted to the National Institute on Diabetes and Digestive and Kidney Disorders (NIDDK). This is one of a series of projects proposed or underway to increase scientific assessment of nonhuman primates in ways that will contribute to the health and well-being of individual primates and other members of their species while providing information that is critically relevant to human public health. The first project of this series is entitled The Great Ape Aging Project (see Erwin *et al.,* this volume). A six-month feasibility study and a two-year implementation grant for that project has been funded by the National Institute on Aging (NIA). That project includes partial support for the oldest captive great apes.

The international nature of the diabetes project can contribute in several important ways. One of the problems in the past has been that few primate veterinarians have seen more than one or two cases of spontaneous diabetes in great apes or other primates. Reports of only a few cases have been published in the scientific and comparative medical literature. Creating a database that includes worldwide input can greatly expand awareness of diabetes and its consequences for great apes and humans. Medical management can become increasingly systematic and refined, and more primates will benefit from a more broad-based program. An understanding of diabetes, as it is expressed in a variety of species biologically similar to humans, can substantially contribute to human public health as well.

Many of the concerns expressed about research involving great apes and other primates focus on the degree to which research is exploitative or painful. Designing research systems that integrate the interests of the great apes with those of humans addresses those reservations. Noninvasive or minimally invasive methods can be used to conduct studies that clearly benefit each great ape that participates as a research subject, as well as other members of the species in captivity. This presents an obligation and opportunity. The *obligation* is to provide for their health and well being and to promote a good quality of life. The *opportunity* is to learn as much as possible from them (and for them) within a context of committed consideration and respect.

ACKNOWLEDGMENTS

The authors are especially grateful to Dr. Biruté Galdikas, Dr. Gary Shapiro, and the Orangutan Foundation International for organizing and enabling our participation in the Great Apes of the World Conference, where a version of this paper was presented. We also appreciate consideration of the proposal for funding of this concept by the National Institute of Diabetes and Digestive and Kidney Diseases (NIDDK). Although our first proposal did not result in support, we will continue to seek support for this and related projects. The project emerged, in part, from work supported by a grant from the USPHS/NIH National Institute on Aging (AG14308) to BIOQUAL, Inc.,

J. Erwin, Principal Investigator. The following individuals provided valuable assistance and encouragement: Dr. N. Bodkin, Dr. F. Coulston, Dr. T. Griffin, J. Harbaugh, S. Harbaugh, Dr. J. Landon, K. Landon, Dr. R. Lee, Dr. N. Lerche, Dr. R. McManamon, Dr. T. Maple, Y. Muskita, M. O'Flaherty, Dr. S. Krasnow Parrish, L. Perkins, and Dr. M. St. Claire.

REFERENCES

Erwin, J., Bloomsmith, M., Boysen, S.T., Hof, P.R., Holloway, R., Lowenstine, L., McManamon, R., Perl, D.P., Young, W., Zihlman, A., this volume.

Hansen, B., 1995, Obesity, diabetes, and insulin resistance: Implications from molecular biology, epidemiology, and experimental studies in humans and animals, *Diabetes Care* 18, Suppl. 2: A2-A9.

Howard, C., 1972, Spontaneous diabetes in *Macaca nigra, Diabetes* 21: 1077 1090.

Howard, C., Fang, T.Y., Southwick, C., Erwin, J., Sugardjito, J., Supriatna, J., Kohlhaas, A., and Lerche, N., 1999, Islet-cell antibodies in Sulawesi macaques, *Am. J. Primatol.* 47: 223-229.

Kemnitz, J., Baker, A., and Shellabarger, W., 1994, Glucose tolerance and insulin levels of captive orangutans. Pp. 250-256 in: (Eds. J. Ogden, L. Perkins, and L. Sheeran), *Proceedings of the International Conference on Orangutans: The Neglected Ape*, San Diego: Zoological Society of San Diego.

Miller, S., 1998, Diabetes: Targeting an old adversary, *Medical Laboratory Observer.* 30(4): 30-41.

Rosenblum, I., Barbolt, T., and Howard, C., 1981, Diabetes mellitus in the chimpanzee (*Pan troglodytes), J. of Med. Primat.* 10: 93-101.

Sugardjito, J., Southwick, C., Supriatna, J., Kohlhaas, A., Baker, S., Erwin, J., Froehlich, J., and Lerche, N., 1989, Population survey of macaques in northern Sulawesi, *Am. J. Primatol.* 18: 285-301.

Chapter 15

STUDIES OF AGE-RELATED NEURONAL PATHOLOGY IN GREAT APES

D.P. Perl[1,2], P.R. Hof[2,3], E.A. Nimchinsky[2,3], and J.M. Erwin[4]

[1]Neuropathology Divison, Mount Sinai School of Medicine, One Gustave L. Levy Place
New York, NY 10029

[2]Arthur M. Fishberg Research Center for Neurobiology, Mount Sinai School of Medicine, New York, NY

[3]Neurobiology of Aging Laboratories, Mount Sinai School of Medicine, New York, NY

[4] Division of Neurobiology, Behavior, and Genetics, BIOQUAL, Inc., Rockville, MD

1.   INTRODUCTION

The age-related neurodegenerative diseases, in particular Alzheimer's disease and Parkinson's disease, represent an increasingly significant public health problem. Indeed, it is currently estimated that approximately 4 million elderly Americans demonstrate progressive failure of cognitive abilities, or dementia, with the overwhelming majority of affected individuals suffering from Alzheimer's disease. In addition, it is estimated that almost one million Americans suffer from Parkinson's disease. Both conditions demonstrate a markedly increasing incidence and prevalence with respect to increasing age. With more individuals surviving to advanced age, these numbers are expected to escalate dramatically over the next several decades. Recognition of the high costs and severe long-term disability associated with the age-related neurodegenerative disorders has served to underscore the importance of research on their etiopathogenesis, as well as the study of brain aging in general.

Through the efforts of numerous scientific laboratories there has been a dramatic increase in the past ten to 15 years in our knowledge of human brain aging and the neurological disorders associated with advanced age. Alzheimer's disease and Parkinson's disease, the two major age-related neurodegenerative diseases, are said to be unique to humans despite observations of age-related declines in cognitive and motor performance in nonhuman primates (Herndon, *et al.*, 1997; Moss, *et al.*, 1997). Such comparative approaches have centered on Old World monkeys, primarily macaques, where progressive cerebral accumulation of beta-amyloid, a requisite lesion in brains of patients with Alzheimer's disease, has been documented, yet does not appear to correlate with cognitive performance (Price, *et al.*, 1994). Recognizing that the macaque life span is considerably shorter that that of humans and that macaques fall short of the neuroanatomical and behavioral complexity of humans, we have turned to the great apes to determine what aspects of human brain aging are shared by our closest relatives in the animal kingdom.

2. EVIDENCE FOR THE DEVELOPMENT OF PATHOLOGIC LESIONS ASSOCIATED WITH AGE-RELATED NEURODEGENERATIVE DISEASES IN GREAT APES

Despite the phylogenetic closeness of humans and the great apes, to date only a scant amount of data are available on age-related changes occurring in the brains of these animals. Whether the neuropathologic changes associated with the age-related disorders of humans occur in great apes also remains incompletely studied. Virtually all of the published reports have investigated the possible presence of neurofibrillary tangles and senile plaques in brains derived from great apes at the end of their maximum life span. The neurofibrillary tangle and senile plaque constitute the two major lesions associated with Alzheimer's disease, and the presence of these lesions represents the means by which neuropathologists confirm the diagnosis of this disorder.

The neurofibrillary tangle consists of dense fibrillary deposits within the neuronal cytoplasm. Such fibrillary deposits consist primarily of an abnormally phosphorylated microtubule-associated protein called *tau* (Lee, *et al.*, 1991). The neurofibrillary tangles have a distinctive appearance upon electron microscopic study, namely, the appearance of a pair of filaments wound in a regular helix, or so-called "paired helical filaments" (Kidd, 1963; Wisniewski, *et al.*, 1976). Such changes have never been described in any primate brain specimens. However, based on available published reports, relatively few specimens derived from great apes of advanced age have been adequately examined.

The senile plaque is the other major neuropathologic lesion associated with Alzheimer's disease. The plaque is a complex lesion with the defining feature of a focal accumulation of a 4 kDa amyloid protein referred to as beta-amyloid, βA4, or Aβ (Masters, *et al.*, 1985). There are a number of different forms of plaques, and the terminology in the literature can be confusing. The neuritic plaque has a central core of βA4 amyloid that is surrounded by a corona of abnormally formed or dystrophic neurites. "Neurites" is a term for neuronal processes (axons and dendrites). These dystrophic neurites stain prominently with a variety of silver impregnation histologic stains, such as the Bielschowski stain, which is also commonly used to identify neurofibrillary tangles. Ultrastructurally, the dystrophic neurites contain dense bodies, membranous profiles, and frequently accumulations of paired-helical filaments. In the periphery of the neuritic plaque one also commonly encounters microglial cells and less frequently, reactive astrocytes.

Particularly through the use of immunohistochemical approaches with antibodies directed against portions of the βA4 molecule, it has been recognized that focal diffuse deposits of amyloid occur in the cortex in the absence of any accompanying dystrophic neurites. Such lesions are referred to as diffuse plaques. Diffuse plaques are commonly encountered in the brains of elderly human subjects and can accumulate in large numbers and occupy a considerable extent of the cerebral cortical ribbon. Nevertheless, the appearance of large numbers of diffuse plaques in well characterized humans does not appear to be associated with any evidence of mental impairment (Crystal, *et al.*, 1988).

The beta amyloid arises from the larger amyloid precursor protein, or APP, a highly conserved transmembrane glycoprotein (Kang, *et al.*, 1987). The beta amyloid is derived from the enzymatic cleavage of two putative secretases (β and γ) that split the amino and carboxyl terminals of the 4kDa segment, which then accumulates (Selkoe, 1994). The carboxyl end cleavage is apparently ragged, with longer forms (42 and 43 amino acids) having a tendency to be deposited within the senile plaques while the shorter form (i.e., consisting of 40 amino acids) tends to be deposited in blood vessels as congophilic angiopathy (Prelli, *et al.*, 1988).

Not only does beta-amyloid deposit in the form of senile plaques, there is also a tendency for the amyloid protein to accumulate in the walls of the cerebral blood vessels. This phenomenon is referred to as congophilic angiopathy. In general, the small arteries and arterioles of the leptomeninges and the cerebral cortex are involved. Prelli, *et al.* (1988) found that the shorter forms of the beta-amyloid polypeptide (the 1-40 amino acid long form) predominates in these deposits. These accumulations do not appear to have any overt effect on the functioning of these vessels, although with severe involvement spontaneous intraparenchymal hemorrhage may occur. Such hemorrhages tend to occur in frontal or occipital white matter and may also be multiple.

Selkoe and colleagues (1987) have reported evidence of deposition of beta-amyloid in the form of plaques in the frontal, temporal, and parietal cortex of a 46-year-old orangutan. These plaques stained with Bielschowski silver stains and showed green-red birefringence under polarized light when stained with Congo red. There was apparently no evidence of dystrophic neuritic changes or *tau*-staining material associated with these plaques, and therefore they are to be considered comparable to human diffuse plaques. These authors did not comment on the extent of plaque formation.

Gearing and coworkers (Gearing, *et al.*, 1994; Gearing, *et al.*, 1996) described the presence of amyloid accumulations in the neuropil and in meningeal and cortical blood vessels of two female chimpanzees aged 56 and 59 years. The brain of a 45-year-old male chimpanzee also examined by these workers apparently failed to reveal any stainable cerebral cortical amyloid deposits. The accumulations in the two older animals were amorphous in nature, also resembling the diffuse plaques seen in humans. Most of the plaques were perivascular in location, and only rare plaques demonstrated evidence of dystrophic neurites.

Finally, morphologic examination of the brains of orangutans that were ten, 28, 31, and 36 years old at death (Gearing, *et al.*, 1997) revealed evidence of sparse beta amyloid immunoreactive plaque-like structures in the brains of the three older animals. Such deposits failed to show a positive reaction with silver impregnation stains. Neurofibrillary tangles were not encountered in any of these brain specimens.

3. OPPORTUNITIES FOR AGE-RELATED STUDIES PROVIDED BY THE GREAT APE AGING PROJECT

Through the Great Ape Aging Project (Erwin, *et al.*, this volume) we have identified the oldest 10% of great apes in the United States. The consortium members provided institutions participating in this program with a detailed protocol for obtaining and preparing brain specimens following the death of great apes under their care. They also provided the necessary supplies and reagents needed to obtain the brain specimen and prepare it in such as way as to maximize its use in age-related research endeavors. The protocol employed by the Great Ape Aging Project is patterned after the one developed and employed in the Mount Sinai School of Medicine brain bank activities for the study of human aging, Alzheimer's disease, Parkinson's disease, and schizophrenia. The Project assumes that the oldest animals have the greatest likelihood of dying and will also provide examples of great ape brain specimens from individual animals of advanced age. It should be noted that this initiative is entirely naturalistic in nature and allows the institutions with the responsibility for the care of the animal to be prepared, upon death, to properly obtain a specimen that is optimally handled in order to maximize its utility for modern neurobiologic research.

Using specimens collected by the Great Ape Aging Project to date, we have begun an investigation of brain aging in chimpanzee, gorilla, and orangutan brains of various ages. In our initial studies we have been particularly interested in investigating whether aging in great apes is associated with neuronal loss. We have directed our attention to a population of neurons that are particularly sensitive to neuronal pathology in association with both in normal aging and in cases of Alzheimer's disease (Morrison and Hof, 1997). The layer II cells of the entorhinal cortex is the neuronal population we have chosen for our initial studies. Our approach to this study has been to employ the technique for cell quantitation referred to as stereology (Coggeshall and Lekan, 1996; Gundersen, *et al.*, 1988a; Gundersen, *et al.*, 1988b; Pakkenberg, *et al.*, 1991; West, *et al.*, 1994). Stereology is an approach that permits one to accurately estimate the total number of objects (neurons, in this instance) within a properly defined structure (in this case, the entorhinal cortex). Using this approach, we are counting the total number of layer II entorhinal neurons in chimpanzee brains derived from both young and old animals. Although these studies are still in their preliminary phases, to date in the chimpanzee brains studied no significant loss of these cells has been identified with respect to increased age at death. These studies have also failed to reveal any significant changes in entorhinal cortical neuronal cell volume. We are also investigating the layer II entorhinal neurons for evidence of neurofibrillary tangle formation.

4.   COMPARATIVE NEUROANATOMY APPROACHES

We have also used specimens collected from the Great Ape Aging Project to investigate aspects of comparative neuroanatomy. We have begun with studies of two unusual neuronal cell types that are present in the anterior cingulate cortex of the human brain. The first cell type is distinguished by its morphology and consists of an elongated neuron that is present within layer Vb of area 24 (Nimchinsky, *et al.*, 1995). It is referred to as the spindle cell and was first described in 1888 by Betz. The other cell type is a more typical pyramidal neuron within layer Va that has the unusual property of containing the calcium-binding protein, calretinin, a protein that is usually found only in nonpyramidal neurons (Nimchinsky, *et al.*, 1997). In humans, both of these cell types are also of increased susceptibility to involvement in Alzheimer's disease. In order to investigate whether these cell populations might be specific to a particular branch of the hominid lineage, samples of the anterior cingulate cortex in tarsier, lemur, bushbaby, owl monkey, squirrel monkey, macaque, baboon, gibbon, orangutan, gorilla, chimpanzee, and bonobo were analyzed. Spindle cells were not found in any of the specimens derived from prosimians, Old World and New World monkeys, or small apes. However, these cells were sparse but definitely present in orangutan specimens, and more abundant in the gorilla and chimpanzee, although not nearly as abundant as in the human. Calretinin-immunoreactive layer Va pyramidal cells were also absent in monkey and gibbon brain specimens. These cells were occasionally found in the orangutan, were more commonly seen in gorillas, and were as abundant in chimpanzee as in the human specimens we have examined. These observations suggest that both these cell types might have evolved around the time of the common ancestor of great apes and humans, which might thus signify a key period for the evolution of this cortical area, at least. These represent two rare examples of a morphologically recognizable neuronal subpopulation that show evidence of having evolved in some, but not all, genera in a family. The cortical area that contains both these cells has been implicated in autonomic regulation as well as vocalization in carnivores and primates. There is also evidence that this area of the brain might have evolved further functions relating to speech and emotion processing in the human. Functional evolution of this cortical area is highly likely in the setting of such cellular specialization, especially of cell types that have since been conserved across genera.

5. CONCLUSIONS

It is clear that there is much to be learned from the investigation of brain aging in the great apes. Given the close genetic relationships of humans and great apes and the apes' ability, especially in captivity, to survive to ages that begin to approach that of humans, the parallels and opportunities for research are numerous. To date only aspects of the pathology of Alzheimer's disease have been demonstrated in the brains of great apes of advanced age, but the number of well-studied animals is quite small. If the great apes fail to develop lesions comparable to that of Alzheimer's disease and Parkinson's disease, then considering the 2-3% differences in genetic material, this would argue for a search for differences in genetic loci of importance to the etiopathogenesis of these two important human diseases. Furthermore, this then represents an important opportunity for the study of neuronal loss and other aspects of normal brain aging in the absence of superimposed lesions associated with these two age-related human diseases. The study of the aging process in the great ape species will provide data of importance to understanding the aging process in these animals themselves. As increasing numbers of great apes survive to advanced age in protected environments, the problems associated with aging will assume increasing importance. All such studies have been hampered by the lack of significant numbers of well prepared specimens from appropriately aged animals. The availability of the collection of specimens being prepared by the Great Ape Aging Project represents an important contribution to those interested in these problems.

ACKNOWLEDGMENTS

These studies are supported in part by grants AG-02210, AG-05138, and AG-14308 from the National Institutes of Health. We also thank Rita Vertesi, Daniel Gimmel, and Fred Robenzadeh for their expert technical assistance.

REFERENCES

Coggeshall, R.E. and Lekan, H.A., 1996, Methods for determining numbers of cells and synapses: A case for more uniform standards of review, *J. Comp. Neurol.* 364: 6-15.
Crystal, H., Dickson, D., Fuld, P., Masur, D., Scott, R., and Mehler, M., 1988, Clinico-pathologic studies in dementia: Non-demented subjects with pathologically confirmed Alzheimer's disease, *Neurology* 38: 1682-1689.

Erwin, J., Bloomsmith, M., Boysen, S.T., Hof, P.R., Holloway, R., Lowenstine, L., McManamon, R., Perl, D.P., Young, W., Zihlman, A., this volume.

Gearing, M., Tigges, J., Mori, H., and Mirra, S.S., 1996, Aß40 is a major form of ß-amyloid in nonhuman primates, *Neurobiol. Aging* 17: 903-908.

Gearing, M., Tigges, J., Mori, H., and Mirra, S.S., 1997, beta-Amyloid (A beta) deposition in the brains of aged orangutans, *Neurobiol. Aging* 18: 139-146.

Gearing, M., Rebeck, G., Hyman, B., Tigges, J., and Mirra, S.S., 1994, Neuropathology and apolipoprotein E profile of aged chimpanzees: Implications for Alzheimer's disease, *Proc. Natl. Acad. Sci. USA* 91: 9382-9386.

Gundersen, H.J., Bendtsen, T.F., Korbo, L., Marcussen, N., Moller, A., Nielsen, K., Nyengaard, J.R., and Pakkenberg, B., 1988a, Some new, simple and efficient stereology methods and their use in pathological research and diagnosis, *APMIS* 96: 379-394.

Gundersen, H.J., Bagger, P., Bendtsen, T.F., Evans, S.M., Korbo, L., Marcussen, N., Moller, A., Nielsen, K., and Pakkenberg, B., 1988b, The new stereology tools: Dissector, fractionator, nucleator and point sampling intercepts and their use in pathological research and diagnosis, *APMIS* 96: 857-881.

Herndon, J.G., Moss, M.B., Rosene, D.L., and Killiany, R.J., 1997, Patterns of cognitive decline in aged rhesus monkeys, *Behav. Brain Res.* 87: 25-34.

Kang, J., Lemaire, H.G., Unterbeck, A., Salbaum, J.M., Masters, C.L., Grzeschik, K.H., Multhaup, G., Beyreuther, K. and Muller Hill, B., 1987, The precursor of Alzheimer's disease amyloid ßA4 protein resembles a cell-surface receptor, *Nature* 325: 733-736.

Kidd, M., 1963, Paired helical filaments in electron microscopy of Alzheimer's disease, *Nature* 197: 192-193.

Lee, V.M.Y., Balin, B.J., Otvos, L., Jr., and Trojanowski, J.Q., 1991, A major subunit of paired helical filaments and derivatized forms of normal tau, *Science* 251: 675-678.

Masters, C.L., Multhaup, G., Simms, G., Pottgiesser, J., Martins, R.N., and Beyreuther, K., 1985, Neuronal origin of a cerebral amyloid: Neurofibrillary tangles of Alzheimer disease contain the same protein as the amyloid of plaque cores and blood, *EMBO J.*, 4(11): 2757-2763.

Morrison, J.H. and Hof, P.R., 1997, Life and death of neurons in the aging brain, *Science* 278: 412-419.

Moss, M.B., Killiany, R.J., Lai, Z.C., Rosene, D.L., and Herndon, J.G., 1997, Recognition memory span in rhesus monkeys of advanced age, *Neurobiol. Aging* 18: 13-19.

Nimchinsky, E.A., Vogt, B.A., Morrison, J.H., and Hof, P.R., 1995, Spindle neurons of the human cingulate cortex, *J. Comp. Neurol.* 355: 27-37.

Nimchinsky, E.A., Vogt, B.A., Morrison, J.H., and Hof, P.R., 1997, Neurofilament and calcium-binding proteins in the human cingulate cortex, *J. Comp. Neurol.* 384: 597-620.

Pakkenberg, B., Moller, A., Gundersen, H.J., Mouritzen Dam, A., and Pakkenberg, H., 1991, The absolute number of nerve cells in substantia nigra in normal subjects and in patients with Parkinson's disease estimated with an unbiased stereological method, *J. Neurol. Neurosurg. Psych.* 54: 30-33.

Prelli, F., Castano, E., Glenner, G.G., and Frangione, B., 1988, Differences between vascular and plaque core amyloid in Alzheimer's disease, *J. Neurochem.* 51: 648-651.

Price, D., Martin, L., Sisodia, S., Walker, L., Voytko, M., Wagster, M., Cork, L., and Koliatsos, V., 1994, The aged nonhuman primate: A model for the behavioral and brain abnormalities occurring in aged humans. Pp. 165-175 in: (Eds. R.D. Terry, R. Katzman, and K. Bick), *Alzheimer's Disease*, New York: Raven Press.

Selkoe, D.J., 1994, Normal and abnormal biology of the ß-amyloid precursor protein, *Ann. Rev. Neurosci.* 17: 489-517.

Selkoe, D.J., Bell, D.S., Podlisny, M.B., Price, D.L., and Cork, L.C., 1987, Conservation of brain amyloid proteins in aged mammals and humans with Alzheimer's disease, *Science* 235: 873-877.

West, M.J., Coleman, P.D., Flood, D.G., and Troncoso, J.C., 1994, Differences in the pattern of hippocampal neuronal loss in normal ageing and Alzheimer's disease, *Lancet* 344: 769-772.

Wisniewski, H.M., Narang, H.K., and Terry, R.D., 1976, Neurofibrillary tangles of paired helical filaments, *J. Neurol. Sci.* 27: 173-181.

Chapter 16

METABOLITES OF OVARIAN HORMONES AND BEHAVIORAL CORRELATES IN CAPTIVE FEMALE BONOBOS (*PAN PANISCUS*)

M.H. Jurke[1,2], L.R. Hagey[1], N.M. Czekala[1], and N.C. Harvey[1]
[1]*Center for Reproduction of Endangered Species, Zoological Society of San Diego, San Diego, CA 92112-0551*

[2]*Department of Anthropology, University of California, San Diego, CA*

1. INTRODUCTION

This is the first report of an ongoing Species Survival Plan (SSP) research project that, among other things, aims to increase our knowledge of reproduction in captive bonobos. With the close collaboration between the Center for Reproduction of Endangered Species (CRES) and the zoos of San Diego, Cincinnati, Columbus, and Milwaukee, this study represents the first of its kind in which hormones, perineal swellings, and behavior are being correlated using a large scale sampling of captive bonobos. The bonobo is perhaps the least studied species among the most endangered of the great apes. The captive breeding population is extremely small: 121 animals worldwide outside of Africa (SSP Meeting in Columbus, 1998). Censuses of the number of bonobos in the wild have been problematic and have resulted in a wide range of estimates (Dupain and Van Elsacker, this volume). Considering the issues of the ongoing bushmeat crisis and habitat destruction, estimates of 10,000 individuals or less in the wild are probably not too low.

The specific aims of this chapter are to:
1. characterize ovarian hormone metabolites found in fecal and urinary samples of females;

2.   increase our knowledge about the relationship of hormones, perineal swellings, and behavior in bonobos; and
3.   investigate reproductive parameters such as female menopause and the adult and adolescent males' distribution of copulations and ejaculations across ovarian cycles.

2.   MATERIAL AND METHODS

Urine and fecal samples of female and male individual bonobos from collaborating institutions are collected almost daily in the morning. Freshly collected samples are frozen immediately and stored at -20°C prior to express shipment in coolers to CRES, where hormone analysis is carried out.

A standardized record sheet is used at each facility with categories filled out by animal caretakers on a daily basis, with information such as sexual swellings, menses, mount, thrusting, intromission, and ejaculation. Three categories for recording swellings of each labia and anus are used: 1=complete detumescence, 2=partial tumescence, 3=complete tumescence, no wrinkles, entire area being taut and does not sway, 3.1=same scoring as 3, but a shine is visible on the labia/anus area (see Furuichi, 1987; Vervaecke, *et al.*, unpublished).

This study is based on a multi-institutional approach, so behavioral data collection is kept as simple as possible in order to increase accuracy, comparability, and inter-observer reliability. Trained observers at each institution collect data on a daily basis using check sheets with 17 behavioral categories. Observation methods include all-occurrences of selected behaviors (approach, leave, follow, number of thrusts) and a modified one/zero scoring that we have found to be the most reliable scoring system across institutions. Focal animals are observed daily for 30 minutes with 30-second intervals.

Ovarian cycle activity is monitored throughout the 15-month study period using estrogen and progesterone profiles and swelling charts. Fecal and urinary samples are analyzed by radioimmunoassays (RIA) in combination with high pressure liquid chromatography (HPLC) and gas chromatography-mass spectrometry (GC-MS). A detailed description of the methodology is provided in Jurke, *et al.* (1998).

2.1   Sample Preparation and Radioimmunoassays

Fecal samples are lyophilized for 72 h, and 0.2 g was extracted using a modified extraction method described by Heistermann, *et al.* (1996). After testing various extraction methods and modifications thereof, we found that extraction with 60% methanol in water yielded the best extraction efficiency (79%).

To control for urine concentration, creatinine was measured in each sample using a Beckman Creatinine Analyzer 2 (Jurke, *et al.*, 1994) prior to hormonal assays. RIAs were performed by transferring 0.01ml of sample into duplicate tubes containing 300μl (estrogen) or 500μl (progestin) of phosphate buffer (PBS, pH 7) and 100μl each of tracer and antibody. After incubation for 14 hrs at 4°C, separation of free and bound antibody was achieved by incubation with 250μl of dextran-coated charcoal for 30 min followed by centrifugation for 15 min at 4°C and decanting into scintillation vials. After addition of 5ml of scintillation cocktail (Ultima Gold, Packard, Groningen, The Netherlands), samples were counted on a Wallac 1400 DSA (see also Jurke, *et al.*, 1997).

2.2 Estrogen Assay

The assay used [6,7-^{3}H(N)]-estrone sulfate (E$_1$S)] as tracer (specific activity: 53Ci/mmol; NEN Life Science Products, Boston, USA) and a rabbit antiserum raised against estrone-sulfate bovine serum albumen (C. J. Munro, University of California, Davis). The working range of the reference standard was 39-5,000pg/tube. Cross-reactivity was 250% with estrone, 100% with E$_1$S, 84.9% with estrone-glucuronide, 4.1% with 17β-estradiol, and 2.7% with estradiol-3-glucuronide.

Assay sensitivity was 21.1pg/tube (calculated as mean pg/tube at 90% B/Bo, n=10). Accuracy was determined as 92.7%±4.9% (mean±SD, n=7). The interassay coefficients of variation at 22% and 60% binding were 8.1% and 7.8%, and estimates of intra-assay variation were 6% and 5.2%, respectively (%SD/Mean, n=10).

2.3 Progestin Assay

The assay used [1,2,6,7-^{3}H(N)]-progesterone as tracer (specific activity: 97Ci/mmol; NEN Life Science Products, Boston, USA) and a monoclonal antibody against 4-P-11-ol-3, 20-dione hemisuccinate:bovine serum albumin (1:32,000; provided by Dr. Jan Roser, University of California, Davis). Crossreactivities have been described by Grieger, *et al.* (1990) and Wasser, *et al.* (1994): 96% 5α-pregnane-3β-ol-20-one; 36% 5α-pregnane-3α-ol-20-one; 7.4% 5β-pregnane-3α-ol-20-one; 4.8% 5β-pregnane-3α, 17α-diol, 20α-one; and <1% pregnane-3-glucuronide (PdG). Progesterone was used for hormone standard with a working range of 7.8-1,000pg/tube.

Assay sensitivity was 11.2pg/tube, and accuracy was 95.8%±4.9%. The interassay coefficients of variation at 16% and 85% binding were 10.6% and 10.3%, and 8.5% and 7.6% intra-assay, respectively.

2.4   NanoESI-MS (nanoelectrospray)

Steroid analysis by mass-spectrometry (MS) has been complicated by the limited capabilities of instruments such as electrospray ionization (ESI) to detect specific charged and uncharged molecules in biological fluids. The coupling of a low-flow (nanoelectrospray) technique to ESI intrumentation has resulted in a huge gain in sensitivity such that steroid sulfates can be detected at the 100 attomols/μl level. In our experiments, the molecular weight of steroid conjugates was determined using a PE SCIEX API-3 (Alberta, Canada) modified with a nanoESI source from Protana A/S (Denmark). Orifice setting was at -115V, with ESI voltage set at -650V. Normal-sized palladium-coated, borosilicate glass capillaries from Protana A/S were used for sample delivery. Analyses were performed at the Scripps Research Institute Mass Spectrometry Laboratory (La Jolla, CA).

3.   RESULTS

Hormonal analyses of urinary and fecal samples in bonobos yielded useful results when compared with sexual swellings. Estrogens that are found in urine are a good indicator for ovarian cyclicity; however, in feces, estrogen measurement did not yield meaningful results (Figure 45). HPLC profiles of fecal samples clearly showed an estrone peak that corresponds well with the HPLC profile of (incompletely) hydrolyzed urine (Figure 46). The early HPLC peak (shown in Figure 46) was collected from the HPLC effluent and further analyzed by nanoEIS-MS. The results indicate that this RIA peak is a mixture, with nanoESI-MS peak identification as follows: m/z 349, estrone-sulfate; m/z 447, estrone-glucuronide; and m/z 469, estrone-glucuronide sodium salt (Figure 47). Since the instrument was optimized for detection of sulfates, it emphasized sulfates at the expense of glucuronides. It appears that the estrone in feces is to some degree masked by other components in the extractant and therefore cannot be reliably monitored using this technique.

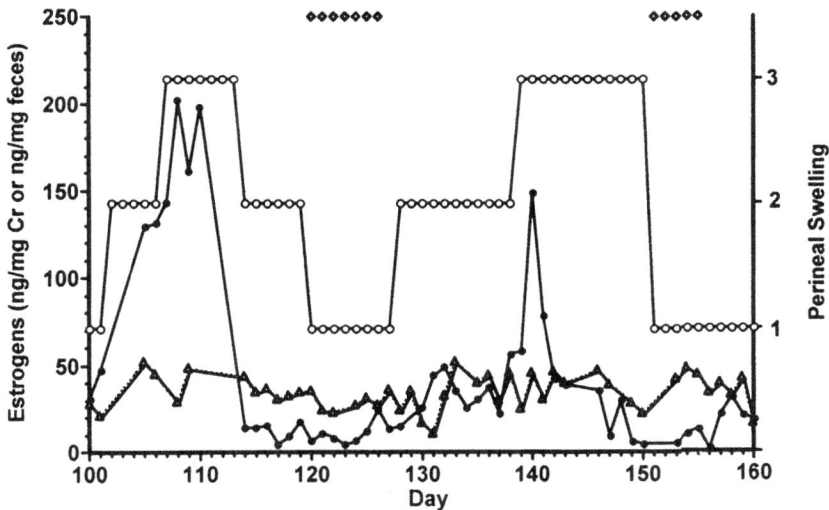

Figure 45. Urinary (●) and fecal (Δ) estrogens across two ovarian cycles. Fecal estrogen measurements seem to be useless for monitoring cycles. ◊: menses; O: perineal swelling.

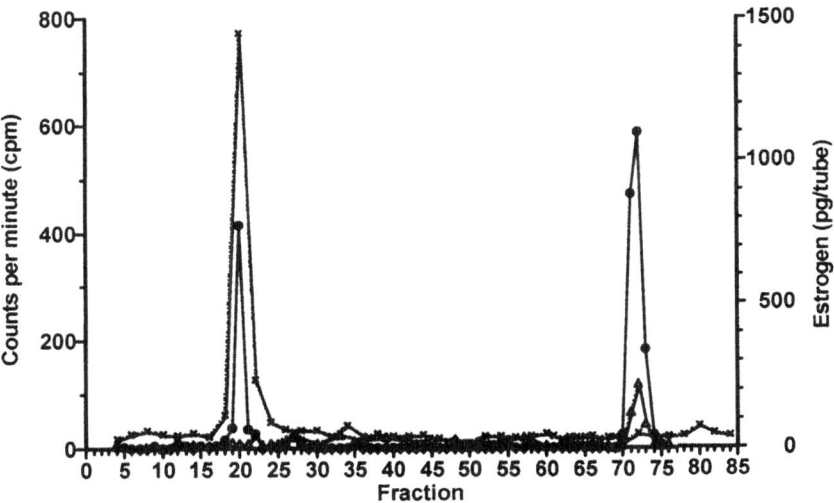

Figure 46. HPLC profiles of a fecal (Δ) and a hydrolyzed urine (●) sample showing peaks at fraction 72 where estrone elutes. x: E₁S tracer.

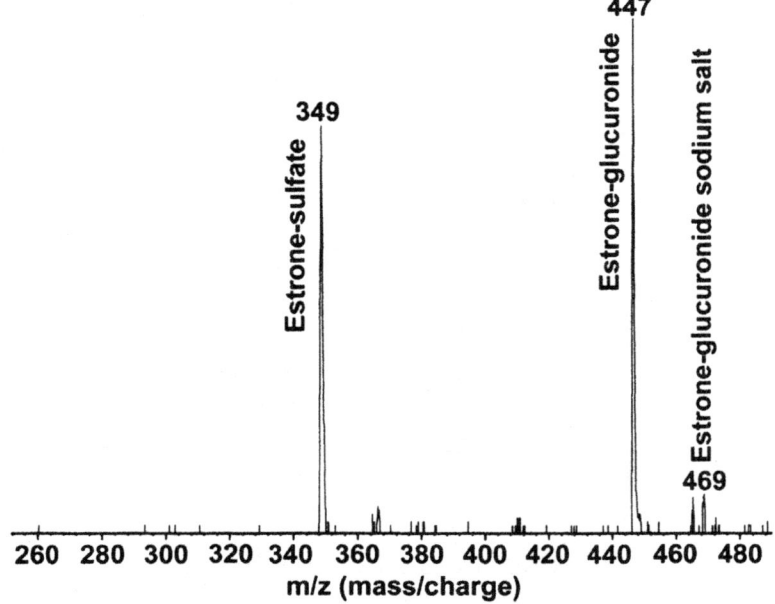

Figure 47. NanoEIS-MS (nanoelectrospray in negative mode) analysis of the early HPLC peak in Figure 46 showing parent compounds of 269 (estrone).

Progestin metabolites in urine appear to be mostly PdG (Figure 48). HPLC profiles of hydrolyzed urine corresponded well with HPLC results of the feces (Figure 49). The free progestin metabolites appear to be primarily from the 5α-pregnane series and were highly correlated with swelling data (Figure 50).

The conspicuous parallelism in the urinary and fecal progestin profiles led us to correlate these two different approaches. Fecal samples yielded approximately ten times the amount of progestins (ng/g feces) compared to urinary progestins (ng/mg creatinine). Progestin metabolite measurements using the same RIA-antibody in 369 pairs of matching urinary and fecal samples, adjusted for delayed excretion in feces by one day (see, e.g., Heistermann, *et al.*, 1996), were used to establish the relationship by applying a simple regression (Figure 51). Out of 19 peaks from 19 partial cycles in four females (female 1: samples 1-125; 2: 126-195; 3: 196-310; 4: 311-369) only the urinary peaks 5, 9, and 10 were not well represented in the feces (Figure 52). These three peaks were also relatively low in urine. Swelling records show that progestins reached a peak around the beginning of detumescence, prior to menses (not shown) in both urine and feces.

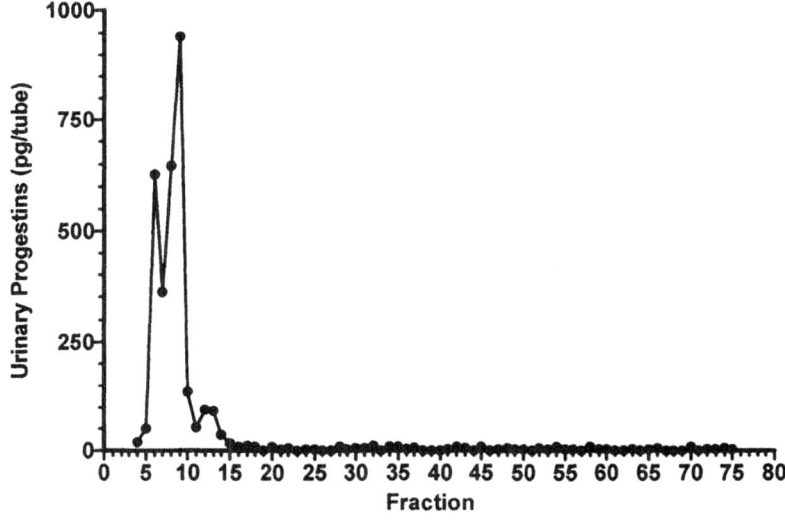

Figure 48. HPLC profile of a urinary sample. The peak at fraction 9 is indicative of PdG.

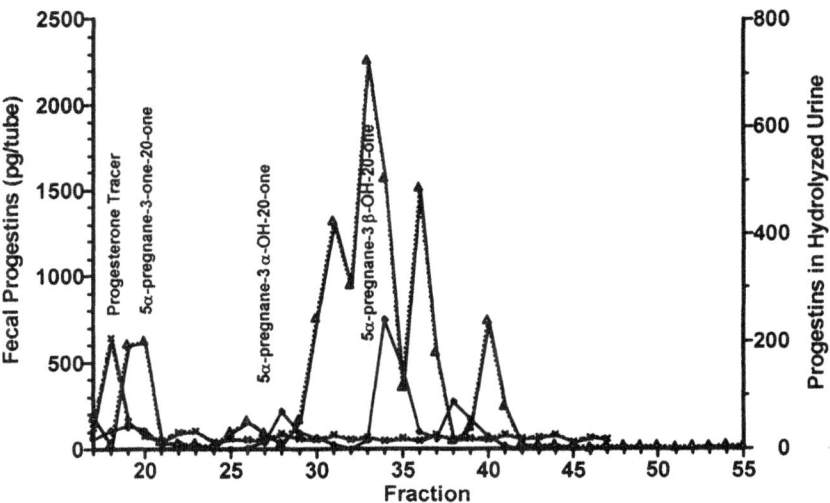

Figure 49. Comparison of HPLC profiles of a fecal and a hydrolyzed urine sample showing peaks indicative of metabolites from the 5α-pregnane series. Same conventions as in Figure 46.

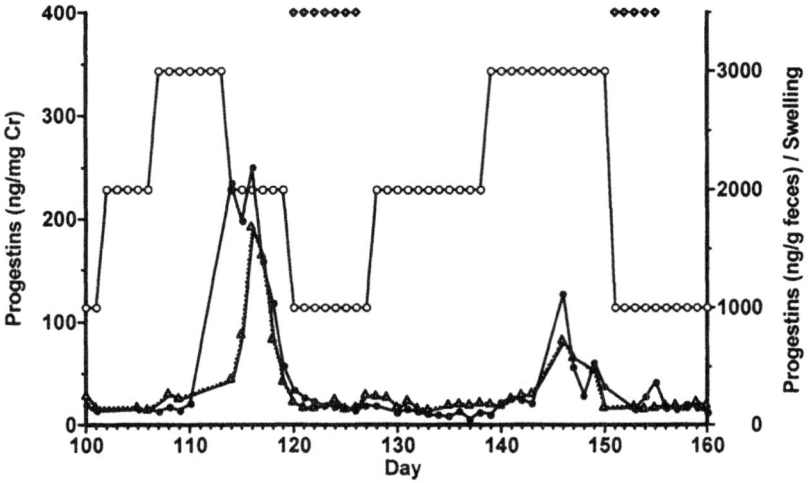

Figure 50. Comparison of fecal and urinary progestins across two ovarian cycles. Both reach a peak at the end of maximum swelling and prior to onset of menses. Same conventions as in Figure 45.

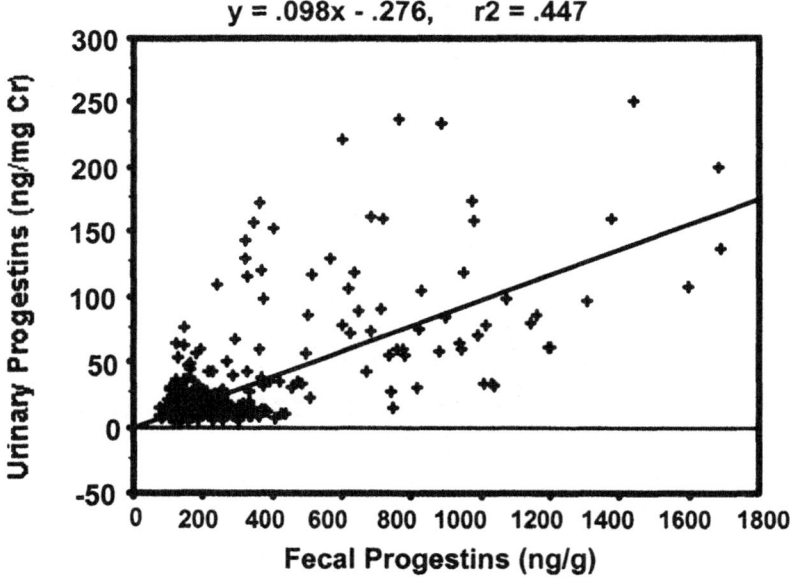

Figure 51. Simple regression of 369 matched pairs of progestin values in urinary and fecal samples.

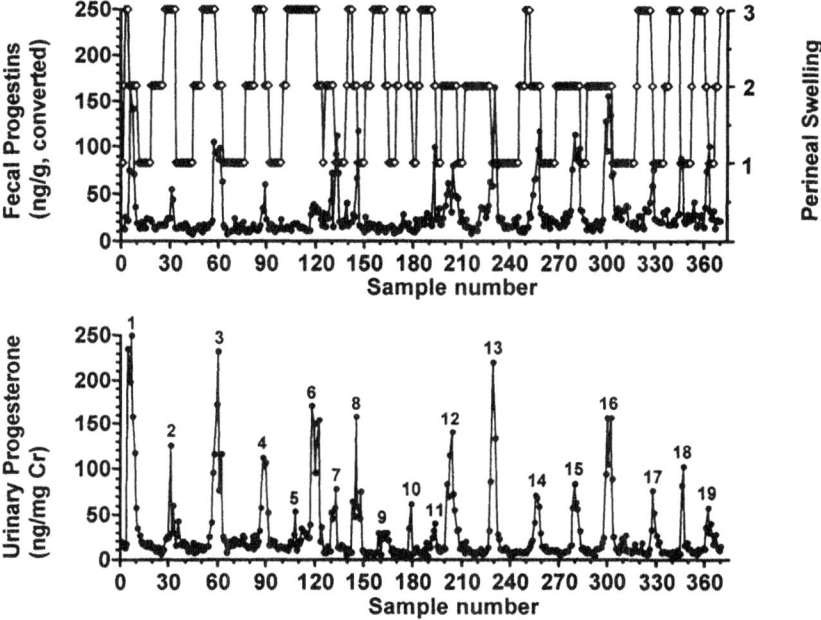

Figure 52. Progesterone metabolites measurements using the same RIA-antibody in 19 partial cycles in four females (female 1: samples 1-125; 2: 126-195; 3: 196-310; 4: 311-369). Fecal samples were converted using the simple regression in (Figure 51) after adjustment for delayed excretion in feces by one day. Swelling records (◊) show that progestins peak around the beginning of detumescence, prior to menses in both urine and feces.

Some of the first results from this long-term study also indicate that the issue of menopause in bonobos is unresolved. The two oldest females, Kitty and Linda, at 47 and 42 years respectively, are still regularly cycling as the following hormonal and swelling profiles demonstrate (Figure 53). Even though Kitty's swelling scores have not reached a 3 or 3.1 for the past year, her hormone levels still show a cycling pattern.

The time of ovulation in bonobos is not obvious to the observer. The maximum swelling of the perineum for extended periods of time masks the precise time of ovulation. However, preliminary behavioral data indicate that adult males tend to copulate (and ejaculate) around the time of peak estrogen, whereas adolescent males are more likely to copulate throughout the female cycle (Figure 54).

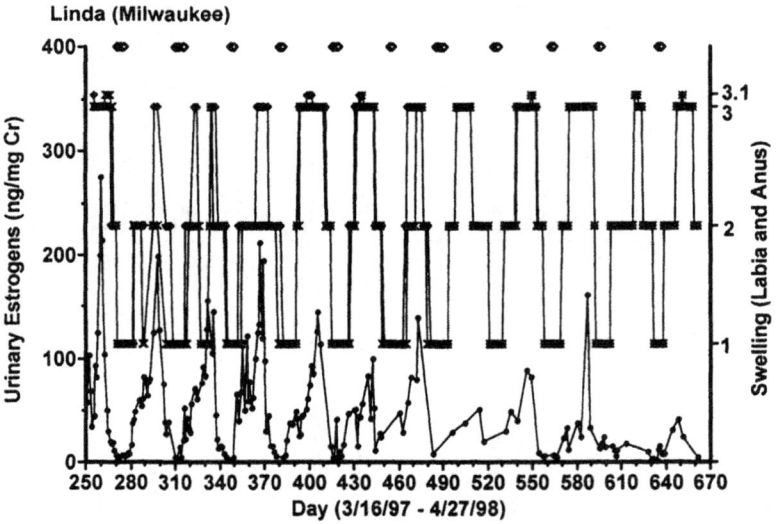

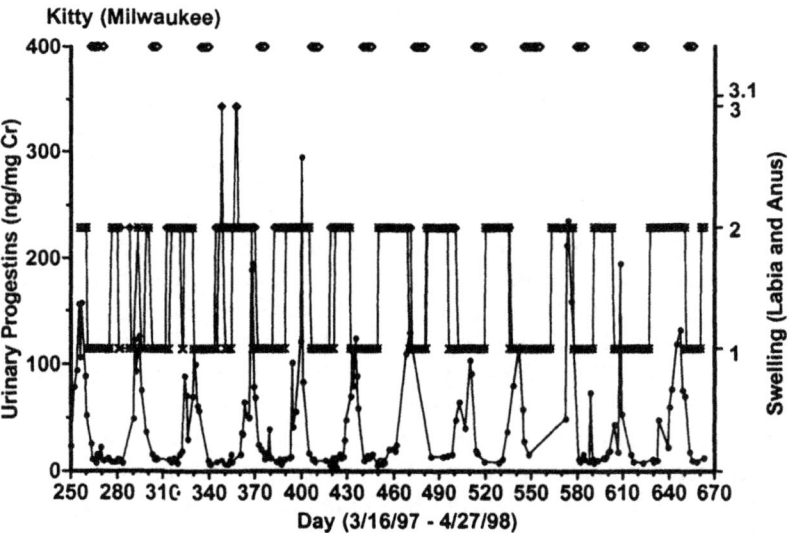

Figure 53. The top graph shows an old female (42) showing regular swelling and hormonal (estrogens) cycling with regular menses (◊). Swellings are differentiated between labial (⊕) and anal (**X**). The bottom graph shows an old female (47) showing reduced degree of swellings and regular hormonal (progestins) cycling with regular menses. Same conventions as in Figure 51.

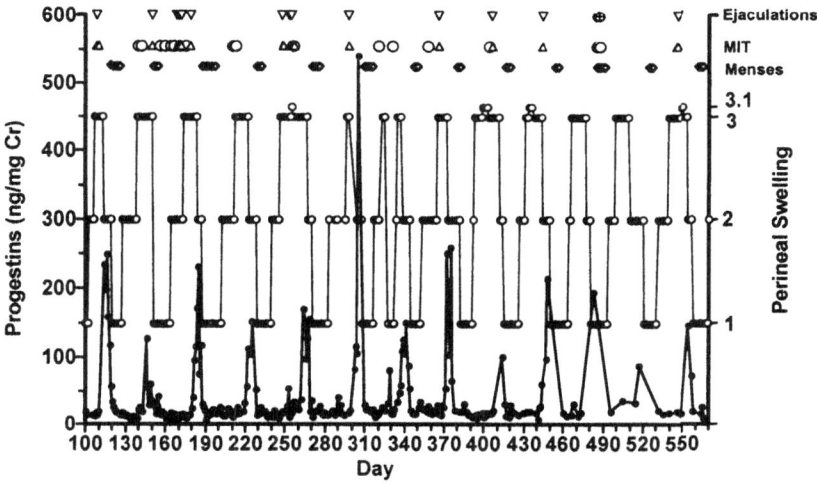

Figure 54. Distribution of adult (Δ) and adolescent (o) copulations defined as mount, intromit, and thrust (MIT) and ejaculations shown on the example of a regularly cycling female across 470 days. Same conventions as in Figure 45.

4. DISCUSSION

Our first results of this large-scale project are consistent with the findings of Heistermann, *et al.* (1996), a study with a much smaller sample size. Our analysis is more extensive and compares fecal and urinary levels of progestins. The standardized approach, which uses the same progestin antibody for both fecal and urine sample analysis, somewhat compromises the comparison of the research efforts by the two research groups. Our antibody crossreacts only marginally with PdG. However, the urinary progestin results are still useful and seem to accurately reflect ovarian activity when correlated with swelling and estrogen data. Given these constraints, conversion of fecal into urinary values as performed here would only be accurate if the same antibody was used. Since all zoological institutions that house bonobo groups in the US are participating in this study conducted at CRES, comparability for bonobos within America is guaranteed. When converted using a regression formula, fecal progestins may complete the picture obtained from urinary progestins, particularly in cases where the urine sample record is unavailable or incomplete.

With regard to menopause in this ape species, we found that perineal swelling alone is misleading. The old females are not yet menopausal. We can dismiss the hypothesis that bonobo females enter menopause at around

45 years of age as has been implied by other researchers without the long-term hormonal data to substantiate their claims. Margrit, the female that was described as menopausal by A. Parish (de Waal, 1997), still shows regular swelling cycles and somewhat irregular menses at 48 years of age (R. Weigel, Frankfurt Zoo, pers. comm.). Without hormonal data, classification as menopausal remains uncertain. In summary, we have yet to find out whether, or at what age, bonobos become infertile, which may have some implications for captive management.

Our preliminary findings also show that adult males tend to copulate at the time of ovulation and peak estrus, while younger males copulate indiscriminately throughout the ovarian cycle. Added analysis of behavioral data in conjunction with hormonal profiles will help to confirm this phenomenon.

5.   CONCLUSIONS

1. Both urinary (predominantly E_1G, E_1S, and PdG) and fecal (mainly metabolites from the 5α-pregnane series and estrone) hormones can be used to monitor ovarian activity in bonobos.
2. Fecal progestin measurements may complete the picture obtained from urinary samples after conversion via simple regression.
3. Fecal estrogen measurements by RIA are unsuitable for monitoring cyclicity.
4. Females of 45 years and older are still regularly cycling. The question of the onset of menopause in this species requires further exploration.
5. Preliminary findings reveal different copulation patterns for adult and adolescent males across the female's ovarian cycle. More behavioral and hormonal data are needed to verify this claim.

ACKNOWLEDGMENTS

This study was partially supported by a Swiss National Science Foundation research grant (823A-046671) and a Conservation and Research grant (Zoological Society of San Diego), both awarded to MHJ. We thank all the animal care providers at the participating institutions for their continuing collaboration in this study. We are also grateful to the many behavioral observers who dedicate countless hours to the behavioral phase of this project. We thank Kelly Chatman, Thomas Holenbeck, and Professor Gary Siuzdak for their time, assistance, and the use of the nanoESI-MS at

Scripps Research Institute. We thank Valentine Lance for his comments on an earlier draft of this chapter.

REFERENCES

Dupain, J., and Van Elsacker, L., this volume.

Furuichi, T., 1987, Sexual swelling, receptivity and grouping of wild pygmy chimpanzee females at Wamba, Zaïre, *Primates* 28(3): 309-318.

Grieger, D.M., Scarborough, R., de Avila, D.M., Johnson, H.E., Reeves, J.J., 1990, Active immunization of beef heifers against luteinizing hormone: III. Evaluation of dose and longevity, *J. Anim. Sci.* 68: 3755-3764.

Heistermann, M., Möhle, U., Vervaecke, H., van Elsacker, L., Hodges, J.K., 1996, Application of urinary and fecal steroid measurements for monitoring ovarian function and pregnancy in the bonobo (*Pan paniscus*) and evaluation of perineal swelling patterns in relation to endocrine events, *Biol. Reprod.* 55: 844-853.

Jurke, M.H., Czekala, N.M., Fitch-Snyder, H., 1997, Non-invasive detection and monitoring of estrus, pregnancy and the postpartum period in pygmy loris (*Nycticebus pygmaeus*), *Am. J. Primatol.* 41: 103-115.

Jurke, M.H., Pryce, C.R., Döbeli, M., Martin, R.D., 1994, Non-invasive detection and monitoring of pregnancy and the postpartum period in Goeldi's monkey (*Callimico goeldii*) using urinary pregnanediol-3α-glucuronide, *Am. J. Primatol.* 34: 319-331.

Jurke, M.H., Czekala, N.M., Jurke, S., Hagey, L.R., Lance, V.A., Conley, A.J., Fitch-Snyder, H., 1998, Monitoring pregnancy in twinning pygmy loris (*Nycticebus pygmaeus*) using fecal estrogen metabolites, *Am. J. Primatol.* 46(2): 173-183.

Vervaecke, H., Van Elsacker, L., Verheyen, R., Heistermann, M., 1998, Unpublished report.

de Waal, F.B.M., 1997, *Bonobo, the Forgotten Ape*, Berkeley: University of California Press.

Wasser, S.K., Monfort, S.L., Southers, J., Wildt, D.E., 1994, Excretion rates and metabolites of oestradiol and progesterone in baboon (*Papio cynocephalus cynocephalus*) faeces, *J. Reprod. Fert.* 101: 213-220.

Chapter 17

SEXUAL MOTIVATION OF MALE CHIMPANZEES DURING THE FEMALE CYCLE, INCLUDING PRELIMINARY DATA ON AGE EFFECTS

R.D. Nadler
Yerkes Regional Primate Research Center, Emory University, Atlanta, GA 30322

1.   INTRODUCTION

Some of the earliest studies of sexual behavior in nonhuman primates revealed that these species mated frequently, irrespective of cycle phase, *when they were tested in conventional laboratory pair-tests* (e.g., Hamilton, 1914; Maslow, 1936). In the case of chimpanzees, it was proposed that the male's influence ("sexual dominance") and female acquiescence, rather than heightened female sexual motivation (estrus), accounted for copulation that was temporally dissociated from ovulation (Yerkes, 1939; Yerkes and Elder, 1936). This hypothesis remained untested for over 40 years. Recent laboratory research on three species of great ape (chimpanzee, gorilla, and orangutan), however, supported the hypothesis (Nadler and Collins, 1991; Nadler, *et al.*, 1994). Copulation temporally dissociated from ovulation was indeed directly related to the male's dominance over the female when the animals were tested in conventional laboratory pair-tests (in which the partners are freely accessible to each other). When the pairs were tested *under conditions in which the female controlled sexual access*, the male's influence was reduced and copulation was more closely associated with ovulation. Similar results were obtained in laboratory studies of group-living chimpanzees, in which the females had greater control over sexual access than they do in conventional pair-tests (Wallis, 1982, 1992). The data from the female-controlled tests and group-living animals supported the hypothesis that the increased frequency and distribution of copulation in the cycle which occurred in the conventional tests was due to the male's influence, i.e., the male's relatively strong and acyclic sexual motivation and assertive initiation of copulation under conditions in which the female's options for refusing, avoiding, or escaping from the male are limited.

One of the more provocative results in the early and more recent laboratory studies of chimpanzee sexual behavior was the difference between them in the extent to which the males initiated copulation dissociated from ovulation. Whereas a considerable amount of copulation of this type was found in the early studies (Yerkes, 1939; Yerkes and Elder, 1936), very little was found in a recent one (Nadler, et al., 1994). Examination of the ages of the males in the two studies suggested a basis for the different results. The males of the first study ranged in age from 7 to 14 years, while those in the recent study ranged from 17 to 40 years. This suggested that the difference in sexual behavior of these animals was a function of age, i.e., it was the younger age of the males in the early studies that accounted for their greater sexual motivation and assertiveness in initiating sexual interactions.

In order to test prospectively the hypothesis that the sexual motivation of male chimpanzees is related to age, a study was conducted with oppositely sexed pairs of chimpanzees that were tested in conventional laboratory pair-tests throughout the female's menstrual cycle. Anogenital swelling of the female was recorded as an indication of cycle phase (and as an indirect estimate of female hormonal state), and penile erection was recorded as a measure of male sexual arousal, one component of male sexual motivation. Measures of male sexual arousability, the other component of male sexual motivation, were derived from the arousal measures (see below).

2.   METHODS

2.1   Subjects

The subjects were eight sexually mature and experienced chimpanzees (*Pan troglodytes*), six males and two females. Four of the males were tested with both females, while two of the males were tested with one (a different one for each). One female was 34 years of age, while the other was 23 years old. The males ranged in age from 7 to 34 years.

2.2   Procedure

The males were tested daily with the same two females during two consecutive cycles of each female, as previously described (Nadler and Bartlett, 1997). For purposes of testing and data collection, the female menstrual cycle was divided into three phases, the early follicular phase (from the day after the first day of menses to one day before the first day of maximal female anogenital swelling), the midcycle phase (from the first to the sixth day of maximal swelling), and the luteal phase (from one day after the first day of detumescence to the first day of menses). Testing was not conducted during the latter days of maximal swelling on order to avoid pregnancy. Each male was tested approximately ten days during the early follicular and luteal phases of the cycle and five days during midcycle. Testing was initiated by opening the door between the male and female outdoor cage compartments 5 min after the male was introduced into his compartment. Tests were terminated 15 min after the door was opened between the male and female compartments.

The following categories of behavior were analyzed: penile erection, rated on a scale of 0-2; 0, the penis is completely retracted; 1, at least 2.5 cm of the penis is extruded beyond the sheath, but the penis is not rigid; 2, the penis is fully extruded and rigid; copulation, intravaginal penile insertion with intromissive thrusting; ejaculation (i.e., the ejaculatory pattern), brief maintenance of the final intromissive thrust accompanied by bodily rigidity and trembling of the thighs and buttocks; male courtship display, bipedal swaggering or stationary squatting and rocking with a penile erection directed toward a female. All categories were scored during each interval of the 15-min test, at 15 second intervals for the first 2 min and at one minute intervals for the last 13 min, resulting in 21 intervals per test. Female anogenital swelling was assessed daily as previously described (Dahl, *et al.*, 1991).

2.3   Data Analysis

An erection score was recorded for each of the 21 intervals of each 15-min test for each male at each phase of the cycle. The mean of these erection scores was calculated at each time interval for all tests at each phase of the cycle, and this mean interval score was termed the interval erection score. The 21 interval erection scores described the presumptive pattern of sexual arousal throughout the course of the tests for each male during each phase of the cycle. Sexual arousability was defined by the Zuckerman index (Zuckerman, 1971), the mean of the 21 interval erection scores at each phase of the cycle, the mean penile erection score (MPES). For those males that were tested with two females, the mean of their two sets of interval erection scores was used to calculate their MPESs. It was hypothesised that male sexual arousability, as reflected by the MPESs, would be highest during midcycle (when female anogenital swelling was maximal) for the males of all ages (H_1), but that arousability would be inversely related to age at all phases of the cycle, i.e., higher for the younger males (H_2).

3.   RESULTS

Table 19 shows the mean penile erection scores (MPESs) obtained for six male chimpanzees of different ages at three phases of the menstrual cycle. For those males that were tested with two females, their patterns of sexual arousal during tests and their MPESs at the different phases of the cycle were comparable for the two females. This suggests that these measures were reasonably stable and consistent for individual males.

Table 19. Sexual arousability of male chimpanzees during the menstrual cycle, as defined by mean penile erection scores, obtained in conventional laboratory pair-tests of sexual behavior with a female.

Male	Age (Years)	Follicular	Cycle Phase Midcycle	Luteal
Conan	7	0.0	1.4	0.0
Rogger	24	0.0	1.7	0.0
Phineas	27	1.4	2.0	0.0
Iyk	30	0.2	1.1	0.6
Jiggs	33	0.3	0.6	0.2
Homer	34	0.0	1.0	0.0

Sexual arousability, as defined by the MPESs, was highest at midcycle for all of the males, supporting H_1. The three younger males had greater MPESs at midcycle than those of the three older males, supporting, in part, H_2. Contrary to H_2, however, there was no clear relationship between the MPESs and age for the early follicular and luteal phases of the cycle. Another unanticipated difference in sexuality of the males was the maintenance of penile erection for prolonged periods of time following ejaculation by the younger males only.

4.   DISCUSSION

The data provide additional support for the use of penile erection as a measure of sexual arousal in male chimpanzees. The derived measure of sexual arousability, the mean penile erection score (MPES), was greatest at midcycle for all the males, consistent with earlier data showing that sexual interactions in this species occur primarily during the midcycle phase of maximal female anogenital swelling (Nadler, *et al.,* 1994; Wallis, 1982, 1992; Yerkes, 1939; Yerkes and Elder, 1936). There was also some support for the hypothesis that sexual arousability is inversely related to age in male chimpanzees; the MPESs during midcycle were greater for the younger than for the older males. It was also found that the three younger males retained penile erection for some time after ejaculation occurred, whereas erection subsided in the older males relatively quickly following ejaculation. It is, perhaps, not surprising that sexual arousability was not clearly related to age at all phases of the cycle, given the relatively small number of males tested thus far, the limited age range of the males (with only one male under the age of 23 years), and the substantial individual differences among chimpanzees of both sexes. As such, idiosyncratic characteristics of either the males or of the females or of the social relationships within individual pairs could have compromised generalities and hypotheses. It was noted, for example, that the youngest male (Conan) initially was not dominant to the older female with which he was tested. Since male dominance over the female was the main factor that permitted males to intimidate and coerce females to mate in earlier studies (e.g., Nadler, *et al.,* 1994), the lack of dominance over the female for the youngest male may well have inhibited him from completely expressing his sexual arousal. A longitudinal analysis of the penile erection data over the course of testing for this male may be informative, since he appeared to become more and more dominant to the female over time. This result also suggests that it is important in studies of this type to pretest male-female pairs for an extended period of time to confirm that a stable social relationship exists within the pairs.

The use of penile erection as a measure of male sexual arousal in chimpanzees was supported, as was the derived measure of sexual arousability. This measure may be usefully applied to any number of species to facilitate interspecies comparisons and comparisons within the same species, such as between laboratory and field studies. The measures also enable the assessment of male sexual motivation in the absence of direct contact with a female, and in response to experimentally varied stimuli of potentially sexual significance, such as photographs and video displays (Sachs, 1995).

ACKNOWLEDGMENTS

The research was supported by USPHS Grant no. RR-00165 from the National Center for Research Resources, NIH, to the Yerkes Center and NSF Grant no. IBN91-09441 to the author.

REFERENCES

Dahl, J.F., Nadler, R.D., and Collins, D.C., 1991, Monitoring the ovarian cycles of *Pan troglodytes* and *P. paniscus*: A comparative approach, *Am. J. Primatol.* 24: 195-209.

Hamilton, G.V., 1914, A study of sexual tendencies in monkeys and baboons, *J. Anim. Behav.* 4: 295-318.

Maslow, A. H., 1936, The role of dominance in the social and sexual behavior of infrahuman primates: I. Observations at Vilas Park Zoo, *J. Gen. Psychol.* 48: 261-277.

Nadler, R.D. and Bartlett, E.S., 1997, Penile erection: A reflection of sexual arousal and arousability in male chimpanzees, *Physiol. Behav.* 61: 425-432.

Nadler, R.D. and Collins, D.C., 1991, Copulatory frequency, urinary pregnanediol and fertility in great apes, *Amer. J. Primatol.* 24: 167-179.

Nadler, R.D., Dahl, J.F., Collins, D.C., and Gould, K.G., 1994, Sexual behavior of chimpanzees (*Pan troglodytes*): Male vs. female regulation, *J. Comp. Psychol.* 108: 58-67.

Sachs, B.D., 1995, Placing erection in context: The reflexogenic-psychogenic dichotomy reconsidered, *Neurosci. Biobehav. Rev.* 19: 211-224.

Wallis, J., 1982, Sexual behavior of captive chimpanzees (*Pan troglodytes*): Pregnant vs. cycling females, *Am. J. Primatol.* 3: 77-88.

Wallis, J., 1992, Chimpanzee genital swelling and its role in the pattern of sociosexual behavior, *Am. J. Primatol.* 28: 101-113.

Yerkes, R. M., 1939, Sexual behavior in the chimpanzee, *Hum. Biol.* 11: 78-111.

Yerkes, R. M. and Elder, J. H., 1936, Oestrus, receptivity and mating in the chimpanzee, *Comp. Psychol. Monogr.* 13: 1-39.

Zuckerman, M., 1971, Physiological measures of sexual arousal in the human, *Psychol. Bull.* 75: 297-329.

SECTION SIX

THE BUSHMEAT CRISIS: AFRICAN APES AT RISK

INTRODUCTION

Volume One closes with three chapters that explore various dimensions of ape conservation, particularly in Africa. The myriad threats to our closest living relatives have reached crisis proportions and were described in earlier chapters in this volume (Boysen and Butynski; Dupain and Van Elsacker; Patterson and Matevia). The urgency expressed by primatologists has been communicated to politicians with some positive results. The US Congress recently proposed the Great Ape Conservation Act, which was signed into law by President Clinton in November 2000 (Public Law Number 106-411). The US Congress notes in the Act's second section that "great ape populations have declined to the point that the long-term survival of the species in the wild is in serious jeopardy" (Great Ape Conservation Act of 2000:1). The Act's authors further describe an array of conservation challenges, including human encroachment into remaining habitat, commercial logging, hunting for the bushmeat trade, population fragmentation, and exposure to disease (Great Ape Conservation Act of 2000:1-2). The authors point out that apes are important flagship species for the habitats they occupy, so hundreds or thousands of species will benefit from their effective protection, including our own species. The Act goes on to earmark funds to conserve and protect all apes. Clearly, now is the time to act on behalf of apes with fresh perspectives that will yield rapid but enduring results.

In Chapter 18, Tony Rose advocates the creation of a conservation-based social movement. He notes that, while data on a species' demography and ecology are important components of conservation plans, these data alone have infrequently moved us toward conservation goals in the broader social arena. Habitat destruction and species' decline are societal problems, and Rose argues that successful solutions must meet humankind's physical, spiritual, and emotional needs. Although the spiritual dimensions of connectedness to other living things and the aesthetic appeal of nature are mentioned in conservation biology texts (e.g. Cowlishaw and Dunbar, 2000; Meffe and Carroll, 1997), primatologists with a few exceptions have been reluctant to join forces with various religious movements that aim to protect the "Creation" and revere all life forms. Rose notes that only such a union will give the conservation movement the moral grounding needed to move forward. On a more pragmatic note, he points out that US citizens donate more money to religious institutions and causes than to any other charity. He urges changes in the language of CITES to reflect the spiritual significance of other life forms. Rose further encourages conservationists to form partnerships with successful businesses, which can help with financial challenges and contribute a unique approach to conservation problems that would likely include increased reliance on internet and other technologies. Rose envisions that new primate conservation leaders will combine "ecology and applied social science" to satisfy human needs while promoting conservation goals.

Throughout the world, the logging industry has increased hunters' access to wildlife, and vertebrates are being slaughtered on an unprecedented scale. John Robinson and his colleagues note "...commercial logging hugely increases the harvest of wildlife by opening up remote forest areas, bringing in people from other regions, and changing local economies and patterns of resource consumption" (Robinson, et al., 1999: 595). They further estimate that in tropical countries of Africa alone, more than one million metric tons of bushmeat are harvested each year (Robinson, et al., 1999: 595). Members of the logging industry must be brought into any effort to halt the bushmeat trade, ideally by prohibiting hunting and the sale of bushmeat in and through their concessions and by providing alternative protein sources for their workers (Robinson, et al., 1999; see also Rose, this volume).

In Chapter 19, Jef Dupain and L. van Elsacker present a case study of the bushmeat trade in the Lomako region of the Democratic Republic of Congo. They describe how logging impacts on subsistence patterns of villagers living nearby and how bushmeat is collected by local and immigrant hunters to sell or to trade for other necessities. Overhunting in one area causes the hunters to move ever deeper into the forest to find food for themselves and their families and products to sell at markets in larger cities such as Kinshasa and Basankusu. Meanwhile, a decline in the local inhabitants' traditional occupations due to political instability and inadequate governmental infrastructure results in a lost market for crops and a dramatically altered way of life. Dupain and his colleagues, working through the Bonobo in Situ Project, have developed programs to decrease the local peoples' reliance on hunting. Project members are also acting as intermediaries between local farmers and businesspeople in Kinshasa to help ensure the sale of various crops. Dupain and his colleagues conducted surveys in market towns to provide more details on dietary preferences and the numbers and types of vertebrates sold. These data will be used to help reduce the demand for bushmeat and other forest products (see also Dupain, et al., 2000).

Although the gorillas and chimpanzees are among the most studied of mammalian species, information about populations throughout the species' geographic ranges are lacking. This is due in part to unstable governments in many African countries and the existence of many as yet unexplored regions. In Chapter 20, Karl Ammann and Nancy Briggs describe the search for an as yet unknown gorilla, *Gorilla gorilla uellensis*. Their story raises the tantalizing possibility of more ape diversity than is currently recognized and is an example of the difficulties one encounters when working in these regions of Africa.

This volume on the African Apes closes, appropriately, with a discussion of the moral status of apes. Paola Cavalieri and Peter Singer elaborate on their earlier argument from *The Great Ape Project* (Cavalieri and Singer, 1995) that apes should be accorded rights that include liberty, life, and freedom from torture. Surely there can be no stronger argument for ape conservation than the observation that apes are, indeed, thinking and feeling beings. This observation is the implicit underpinning of the US Congress' Great Ape Act and is abundantly supported by field and captive studies of the apes, including data presented in several chapters in this volume.

REFERENCES

Boysen, S. and Butynski, T., this volume.

Cavalieri, P. and Singer, P. (eds.), 1995, *The Great Ape Project: Equality Beyond Humanity*. New York: St. Martin's Press.

Cowlishaw, G. and Dunbar, R.I.M., 2000, *Primate Conservation Biology*, Chicago: University of Chicago Press.

Dupain, J. and Van Elsacker, L., this volume.

Dupain, J., Van Krunkelsven, E., Van Elsacker, L., and Verheyen, R.F., 2000, Current status of the bonobo (*Pan paniscus*) in the proposed Lomako Reserve, Democratic Republic of Congo, *Biological Conservation* 94: 254-272.

Meffe, G.K. and Carroll, C.R., 1997, *Principles of Conservation Biology*, 2nd edition, Sunderland, MA: Sinauer.

Patterson, F.P.G. and Matevia, M.L., this volume.

Robinson, J.G., Redford, K.H., and Bennett, E.L., 1999, Wildlife harvest in logged tropical forests, *Science* 284: 595-596.

Rose, A.L., this volume

US Congress, Great Ape Conservation Act of 2000, 106th Congress of the United States of America, Public Law 106-411, November 1, 2000.

Chapter 18

BUSHMEAT, PRIMATE KINSHIP, AND THE GLOBAL CONSERVATION MOVEMENT

A.L. Rose
The Biosynergy Institute, P.O. Box 488, Hermosa Beach, CA 90254

1.   INTRODUCTION

It is said in theory and affirmed by research that the major revolutions in any science, and in most pursuits of knowledge, are set off primarily by players who come to an issue from outside the traditional domains of its study. In this position paper that premise is put to work. Primatologists from laboratory to field station have focused on nonhumans, trading on their understanding of apes and monkeys to postulate theories of mind, tenets of social behavior, and principles of biodiversity. The opposite is done here. In this treatise I describe findings and theories in human values, behavior, and social systems as directed toward conservation, the penultimate challenge to biological disciplines. Few will contest the fact that conservation is a decidedly human affair, and that its problems and practices have more to do with clothed people than hairy animals. Here I explore the human factors that are transforming conservation from a narrow biological endeavor to a massive social movement. Inter-disciplinary thinking is fundamental to the social movement that wildlife conservation must become in the era of bushmeat and primate kinship.

2.   THE BUSHMEAT CRISIS DOMINATES
AFRICAN PRIMATE CONSERVATION

Across the forest regions of west and central Africa, a confluence of factors are making human predation a leading threat to the survival of many primates, including the great apes. Primate hunting is reported in 27 of the 44 primate study and conservation projects described in the World Conservation Union's (IUCN) recent status survey on African Primates (Oates, 1996a). In twelve of these territories, human predation is a severe threat to species survival. The 2000 IUCN *Red List of Threatened Species* shows a jump in the numbers for mammals, with the order primates most threatened by extinction. The situation is worse in those areas where most remaining apes and monkeys live, outside parks and reserves. In Africa, hundreds of unique and never studied primate populations are being annihilated, and thousands will follow if the current trends continue (Ammann, 1998a; Oates, 1996b; Rose, 1996a).

The risk level for different populations and species varies with their numbers, reproductive vigor, and geographic distribution. Declines in the past have been correlated most closely with human population growth and the destruction of habitat. Primate hunting, including apes, has long been recognized as a factor. Eltringham (1984:34) wrote that "Gorillas and chimps costing several thousand dollars each are captured for zoos and medical research centers, but the quantity killed for food dwarfs the number taken alive." While capture of live apes for research has mostly stopped, a growing body of evidence now shows that shifts in human social and economic practices in the forests of Africa have greatly increased killing for meat. Oates (1996b:8) concludes "while the total removal of natural habitat is clearly a major threat to the survival of many African forest primates, an analysis of survey data suggests that human predation tends to have a greater negative impact on primate populations than does selective logging or low-intensity bush-fallow agriculture."

Ammann's (1993, 1996a, b) wide ranging investigation of hunting pressures in and outside IUCN-surveyed project areas strongly indicates that unprotected and unstudied groups of primates—especially those within 30 km of the expanding network of logging roads and towns—are being devastated by a burgeoning commercial bushmeat trade. The catalyst of this devastation is growth of the timber industry (Ammann, 1996c; Ammann and Pearce, 1995; Dupain and Van Elsacker, this volume; Thompson, this volume).

Timber prices and profits are tied to provision of subsidized bushmeat to migrant workers. Every logging town has its modern hunting camp, supplied with European-made guns, internationally-made ammunition, and men and women who come from towns and cities hoping to make a living in the forests. With indigenous forest dwellers hired as guides and hand servants, immigrant hunters comb the forests, shooting and trapping. Anything edible is fair game in a market that starts with the wood cutters, truck drivers, and camp families who scrape together their meager wages for scarce protein. From this captive market base the bushmeat trade stretches all the way to fine restaurants and private feasts in national capitals where more rare and expensive fare is available. Little is done to teach or enforce wildlife laws. Giant pangolin (*Manis*), gorilla (*Gorilla*), chimpanzee (*Pan troglodytes*), and elephant (*Loxodonta*) are among the animals that are slaughtered in timber concessions and sold for their meat. This scenario is so pervasive, and so driven by human values and economics, that it is the rule wherever logging roads and buildup of timber company personnel occur in the forests.

Most timber executives admit there is a problem and say they are powerless to stop it (Incha, 1996; Splaney, 1998). In the past, logging managers have been reluctant to let outsiders into their concessions, fearing that problems will be uncovered and business disrupted, with no solutions provided. The timber industry's reliance on bushmeat to feed loggers and their inability to educate workers and govern their concessions leads to indiscriminate hunting that not only fosters the breaking of laws, but also the breaking of customs. People whose colonial and tribal cultures once enforced taboos against eating apes and monkeys are beginning to try it (Ammann, 1998b).

Even in areas with no logging intrusion, growing demand for chimpanzee and gorilla meat can be substantial. Kano and Asato (1994) compared ape density and hunting pressure from 29 Aka and Bantu villages along the Motaba River area of northeastern Congo Republic and projected a bleak future for the apes. They found that over 80% of their 173 Aka informants were willing to eat gorilla or chimpanzee meat. Among 120 Bantu informants, 70% were willing to eat gorilla meat and 57% would eat chimpanzee. Because more Aka were involved in ape hunting, 40% reported having eaten gorilla or chimpanzee meat in the previous year, while 27% of Bantu had eaten apes in the same period. Aka informants estimated 34 to 60 successful "subsistence" hunters slaughtered 49 gorillas and 103 chimpanzees in 1992. Bantu informants claimed seven to nine hunters killed 13 gorillas and 28 chimpanzees that year. Kano and Asato (1994: 161) measured ape population density and assert that the survival of both ape populations is at serious risk in this territory, as it is further east for the bonobo, "unless a strong system can be established which combines effective protection with the provision of attractive substitutes for ape meat to the local people." The finding that village hunting of apes in a large habitat area is unsustainable when guns are used makes us all the more concerned about the popular and organized commercial bushmeat trade supported by timber industry infrastructure that is feeding and fostering consumer preferences in towns and cities.

South of the Motaba River, Hennessey (1995) studied bushmeat commerce around the Congolese city of Ouesso. He reports that 64% of the bushmeat in Ouesso comes from an 80 km road traveling southwest to a village called Liouesso. There a hunter who specializes in apes was responsible for most of the 1.6 gorilla carcasses sold each week in the Ouesso marketplace, over 80 gorillas per year in one city. Hennessey projects that 50 forest elephants and 19 chimpanzees are killed annually.

Similar Aka-Bantu hunting and long-distance commercial bushmeat trade is described by Wilke, et al. (1992) in the Sangha region west of Ouesso. There, many hunters preferred trading their meat at Ouesso in order to get a higher price than at logging concessions, confirming the report of Stromayer and Ekobo (1991) that Ouesso and Brazzaville are the ultimate sources of demand. Wilke, et al. (1992) describe monkey meat for sale, but say nothing about apes. They recommend that wildlife conservation officers and biologists monitor and protect duiker, primates, and elephants to regulate "the harvest of forest protein."

Ammann and Pearce (1995) report intense hunting of apes for bushmeat in southeastern Cameroon, across the border west of Wilke's study site. "The hunters in the Kika, Moloundou and Mabale triangle in Cameroon estimate that around 25 guns are active on any given day and that successful gorilla hunts take place on about 10% of outings. This would result in an estimated kill of up to 800 gorillas a year (Ammann and Pearce, 1995: 13)." These same hunters kill up to 400 chimpanzees per year. While some of this ape meat is sold to logging workers in these forests, most is shipped on logging lorries back to Bertoua and all the way to Yaounde and Douala where a better profit can be made. Ammann (1998a) has confirmed Hennessey's (1995) findings that a small portion of Cameroon bushmeat crosses the border for sale in Ouesso.

Illegal bushmeat including gorilla, chimpanzee, and bonobo in villages near reserves like Lope, Ndoki, and Dja, and in city markets at Yaounde, Bangui, Kinshasa, Pt. Noire, and Libraville, has been photographed by Ammann (1996a, 1997, 1998b; McRae and Ammann, 1997). Traders interviewed in those areas affirm that the fresh meat comes from nearby forests, while smoked viand can travel long distances. The scant million people who inhabit the large forested territory of Gabon have a strong palate for bushmeat. Steel (1994) found half the meat sold in Gabon city markets is bushmeat, an estimated $50 million unpoliced trade. Primates comprise 20% of the bushmeat. This includes some apes, which are considered edible by various local tribes. Absent region-wide monitoring of hunting and bushmeat trade, one can only guess the numbers of primates killed to feed the tens of millions of people living in equatorial Africa. There can be little doubt that many more apes are butchered for meat in the lowland forests every year than live captive in all the world's zoos, laboratories, and sanctuaries; perhaps 3,000-10,000 a year!

During extensive discussion with field researchers and conservationists (Rose, 1996b, c, d, e; Rose and Ammann, 1996), I found expert consensus predicting that "if the present trend in forest exploitation continues without a radical shift in our approach to conservation, most edible wildlife in the equatorial forests of Africa will be butchered before the viable habitat is torn down" (Rose, 1996e: 1). Even more worrisome is the agreement among primatologists that the varied destructive outcomes of bushmeat commerce have reached crisis proportions (Rose, 1996b). Juste, *et al.* (1995: 465) crystallize the essence of the crisis: "With the advent of modern firearms, and improved communications and transport, subsistence hunting has given way to anarchic exploitation of wildlife to supply the rapidly growing cities with game."

The key word here is *anarchic*. Absent an effective political authority, having no cohesive principle, common standard, or purpose, the bushmeat trade has exploded into a rush for personal profit not unlike the gold rush that transformed the western portion of the United States in the last century. One timber company executive described it rhetorically: "if you found this hundred franc note lying on the ground, would you pick it up?" (Incha, 1996).

Bushmeat commerce grows with the logging industry, but it is founded on the complex cultures of the region. When people see an animal as little more than meat, they will hunt, butcher, and eat it with impunity (Cartmill, 1993). Mittermeier (1987) warned of the pervasive global threat of primate hunting over a decade ago. Goodall (1998: 7) declared that "unless we work together to change attitudes at all levels—from world leaders to the consumers of illegal bushmeat——there will be no viable populations of great apes in the wild within 50 years." The day will come when all the logging and transport roads are built, the choice wood is removed, and the migrant hunters have harvested the bushmeat in the 90% of African rain forests that are targeted for exploitation. Then parks and reserves will be the only places left to hunt: they will need to be defended by armies, or abandoned. Just as profiteers seek the last black rhino horn in Zambia, so will trophy hunters attempt to buy the last gorilla loin and chimpanzee arm in the Congo basin.

This destruction is not inevitable. There are opportunities to stop the slaughter of primates and re-engender the reverence for wildlife that will save the natural heritage of Africa. To capitalize on these options, one must expand one's visions, strategies, and tactics and break free of the narrow ideological biases that still control the traditional field of conservation biology.

3.   HUMAN KINSHIP WITH GREAT APES RAISES THE STAKES

The view of apes and other nonhuman primates is changing radically outside of Africa. In developed countries, especially among the more educated, a rising sense of kinship with apes and monkeys is almost palpable. Primatologists have seen many primates exhibit elaborate and exquisite gentility, intelligence, and grace, as well as humor, affection, cunning, and some familiar forms of cruelty and sloth (eg: Cheney and Seyfarth, 1992; Fouts and Mills, 1997; King, 1995; Patterson and Linden, 1981; Savage-Rumbaugh and Lewin, 1996; de Waal, 1990; Wrangham and Peterson, 1996). Biologists have uncovered evidence of close genetic kinship between humans and apes. Primate studies are making inroads into fields that were traditionally human focused, such as politics, law, and ethics (Allen and Bekoff, 1997, Cavalieri and Singer, 1993; Singer and Cavalieri, this volume; de Waal, 1996). These discoveries are being told through magazines and books, television and cinema, and in daily newsprint to an international audience.

The explosion of media and entertainment industry interest in nonhuman primates reflects a deep fascination with our primate heritage. The Vatican has softened its position on evolution, calling it "a hypothesis to consider" and many people around the world are now able to think of themselves as "the third chimpanzee." Television programmers have made wildlife and nature documentaries a mainstay of many people's evening entertainment fare, and the apes are featured most often. These developments ease our crossing of the chasm between ape and human, help people build personal and intellectual bonds with apes and strengthen the impetus to preserve and protect all wildlife.

My research on natural epiphanies (Rose, 1994, 1996a, 1998b) adds to a large body of evidence and belief that humans are endowed with an innate fascination and need to relate to other living beings (Kellert and Wilson, 1993). E. O. Wilson (1984) called this drive "biophilia". Overall, people are most affected and inspired by direct interaction with animals. Communion with nature changes minds and action, but to a lesser degree. Scientific study is a prime mover in relatively few people's lives (Kellert, 1996; Rose, 1994).

Among wildlife professionals and lay people in North America and Europe, a growing constituency is making the crucial shift from *concern about* other primates to the more enduring position of *identity with* them and their plight.   These people's stake in primate conservation is personal, holistic, and expansive.   Many millions consider great apes as kin.   They judge the killing of chimpanzees and gorillas to be murder, and eating them to be cannibalism.   We cannot ignore this potent group, nor should we.   The developed world's new sense of kinship with great apes raises the stakes: it demands that those who conserve wild primates do so for all apes and monkeys, not just for the few who are fortunate to live in favorite parks and reserves.

Ironically, the human values and attitudes that support the bushmeat commerce come from maladaptation of old-style colonial worldviews.   In much of central Africa "a general pattern of apathy, fatalism, and materialism towards nature and wildlife" prevails (Kellert, 1996: 149).   Contemporary Africans have lost their traditional "theistic" reverence for wildlife and have assumed from developed countries a harsh, utilitarian view (Mordi, 1991).   With the advent of cash economy, colonial religion, and central government, "tribal values of conserving and protecting nonhuman life are rendered spiritually inoperable, while new ecological and ethical foundations for sustaining nature have not emerged" (Kellert, 1996: 152).

Wherever traditional theistic values are dead and buried, the most viable shifts in attitude for most Africans will come from instilling humanistic views of wildlife like those emerging in the North.   I am not proposing eco-imperialism that foists colonial Northern values on traditional Africans.   The new Northern sense of kinship with other primates is closer to traditional tribal views than the imported colonial dominionistic values now holding sway in Africa.

In territories where primate eating is not taboo, the people who refuse to eat them do so "because they are too much like us" (Hennessey, 1995; Kano and Asato, 1994).   This identity with primates offers a foundation on which to reconstruct an African conservation ethos that once again reveres wildlife and wilderness, and views humanity as an integral part of the natural order.   I have seen the potential for such change among bushmeat hunters in Cameroon's eastern province (Rose, 1997, 1998a, c).   Most of the people in the bushmeat trade readily say that commercial hunting is a poor way to make a living—not a sought after career but a last choice.   Economic factors are less enduring and shift more quickly than personal values, beliefs, and taboos (Dupain and Van Elsacker, this volume).

Psychosocial development spans a hierarchy of human needs (Maslow, 1993).   Conservationists satisfy the most basic of those needs when they employ local people to protect endangered animals from their neighbors who would kill and butcher them, pitting one *survival* tactic against another (e.g. Owens and Owens, 1992). *Security* for these conflicting factions hinges on the relative stability of two industries—bushmeat and conservation. Some people gain *status* by protecting live primates *in situ* or caring for them in captivity. Others are valued for their ability to track and bring ape and monkey meat back for the cooking pot. But status gained from commercial hunting is low in much of Africa, as is the income. Most hunters are not licensed and thus operate in gray market circles; some are admired for their daring and endurance, but not for much else. People who succeed in primate protection and husbandry are better reimbursed than poachers and meat traders. They are also better accepted in most quarters, especially among the more educated professionals. This begins the expansion of *self identity* required to assure the shift from poacher to protector (Rose, 2001a, 1998a, c).

As in developed countries, Africans who come to identify with fellow primates undergo shifts in values towards nature and expansion of worldview. School children and adults in Cameroon demonstrate increased empathy for apes and concern for their welfare after reading about Koko, the signing gorilla (Rose, 2000). Conservation values education affects people of all ages, causing reconsideration of old myths and of newer colonial precepts. Expanding the ego to include something of primate nature is an impetus to seek greater *self and social actualization*. This is when things really begin to change.

4.   PRIMATE CONSERVATION MUST BECOME A GLOBAL SOCIAL MOVEMENT

Psychologists who study the self-fulfilled end of humanity report that human potential is realized most in those who serve others (eg: Rogers, 1961; Rose and Auw, 1974). It seems that we become more of ourselves when we are less selfish. We become bigger, inclusive, multi-faceted, personally enriched by the act of giving. When people gather and organize into groups to realize their potential altruistically, we have the rudiments of a social movement. Conservation is a massive global social movement. As detailed above, people are attracted to this movement first for interpersonal reasons. They expect conservation to provide them with a deep connection to animals and with personal actualization from doing benevolent service. With this impetus, the conservation movement will not only protect nature; it will change human myth, ritual, and institutions.

In much of the world, myth is now created on film and video and transmitted by the commercial entertainment industry. The public is attuned to the attractive power of fame; leaders of social movements cannot be fully effective without name and face recognition. The conservation movement is no exception. Conservationists will gain more support for their cause smiling at TV cameras than staring into microscopes. Most people are induced to spend their hard earned savings and donate their valuable time by good stories, not good statistics.

4.1   Call for New Leadership

The film *Gorillas in the Mist* is shown in schoolrooms across America promoting Dian Fossey as martyr to the cause. Jane Goodall's impressive persona is revered by millions of people as the "Mother Theresa of chimpanzees." Biruté Galdikas is portrayed as the angel of the orangutan. These women are public icons, but they and their torch-bearers must continue to celebrate and support others who conserve and protect primates and their habitat. This social movement needs scores of conservationists with celebrity status, not two or three.

The opportunity exists. Films and popular books about primatologists and the life ways of nonhuman primates are beginning to proliferate in the Northern marketplace, which reflects and raises public concern and motivates the public to seek more personal means of connecting with primate kin.

Some traditional biologists argue that human contact renders the animals useless to science, puts them at risk of disease, and should be avoided at all costs. Some risks are real, but they must be conquered to keep up with the social movement. Others claim that humanistic approaches present a false view of primates, foster dangerous anthropomorphism, and impede practical scientific work. Ironically there is no scientific evidence to back these arguments. Scientific methods are made to test hypotheses, not conserve wildlife. It is time to recognize that those who measure environmental destruction are not often prepared or able to do much about it.

The conservation movement is a very human affair. It is not just about little groups of hairy animals. It is about hungry humans whose greed and ignorance are putting all of life at great risk. Big changes based on big visions are needed along with the aid of the most wise and benevolent and the most wealthy and powerful people and agencies in the world. Leaders of the conservation movement will come from three forces: business, religion, and ecosocial practice.

4.2   Business Endows

Imagine what it would be like if Bill Gates sponsored the IUCN Primate Specialist Group. A Microsoft approach to conservation would invent new ways to market primates *in situ*—virtual ecotourism, interactive ecology games in real time, mobile distance learning units linked to school wildlife labs. Web surfers could see their favorite animals and conservationists in action, buy a bushmeat-free meal for a hungry park ranger, adopt an orphan ape, save a monkey troop, protect a forest, give advice to policy makers, all on-line. Wealthy zoos would have direct banking links to wildlife sanctuaries in habitat countries. Foresters in developed countries would connect on-line with tropical forest officers to share ideas and resources. Interactive networks of financiers, local community members, wildlife law enforcers, exploiters, religious leaders, and scores of other stakeholders would foster a positive global outlook linking the varied elements of the conservation movement.

If Gates and Mittermeier became partners, people would send ape and monkey holograms to their loved ones for Earth Day via the internet and donate the profit to conservation programs. This entrepreneurial approach is a far cry from that of traditional conservationists with besieged island outlooks seeking little more than protection of their favorite primate study populations. Academia promotes good analysis, but business endows effective action.

The full force of the international business community can capitalize on the social movement and make living wildlife and wilderness more profitable than cut wood and butchered animals. But there are other forces that will take the social movement even farther. Business-like conservation (Ammann, 1996d) is necessary but not sufficient to assure the restoration of African wildlife and wilderness. Entrepreneurs who devise business projects and managers who pursue measured objectives may deliver profit without protection, sanctuary without well being. To pursue altruistic goals that assure humane outcomes requires *moral leadership*. To integrate the needs and capacities of diverse human and natural stakeholders into successful programs that produce synergistic results requires *ecosocial competence*. Both these fundamental imperatives must be brought to bear on the conservation movement. Well endowed action is not necessarily right action (Hawley, 1993).

4.3   Religion Inspires

Leaders of the major religions are organizing and acting on behalf of the environment. The Christian Environmental Council in North America has used Bible citations to challenge corporate environmental ethics and to provide authority for proactive positions on crucial elements of ecological justice—endangered species protection, environmental precedence over private property claims, and control of global climate change (Alexander, 1998). The U.S. Catholic Conference launched its Environmental Justice Program in 1993 and with the new edicts of the Pope, may expand it into their churches and parishes worldwide. Inter-faith groups are proliferating with projects to foster ecological renewal, responding to and amplifying the global call for "love and care for the Creation" (Rose-Erejon, 1998).

To put the potential impact of these developments in perspective, it is important to note that over half the charity dollars spent in the USA go to religious groups, compared to less than 2% for conservation organizations. Donations to religious institutions for benevolent stewardship of the natural world will outstrip the current level of gifts to secular environmental NGOs many-fold in the next decade. More important than money alone will be the deep commitment of billions of people whose concern for nature will have a fresh and enduring outlet. Religious groups enter the conservation arena with double motivation: after the humanistic attachment to animals as kin, the second most prevalent value towards nature in the North is the *moralistic*, which encompasses "strong feelings of affinity, ethical responsibility, and even reverence for the natural world" (Kellert, 1993: 53).

This will be a crucial balance to the business force, but the expansion of religious concern for the natural world warrants support for other reasons besides the balance it provides to business. Perhaps the optimum use of wildlife and wilderness is the religious and spiritual use—to love and care for the natural environment (the Creation) by prayer, meditation, and altruistic service is about as synergistic and sustainable an involvement as one can imagine. To establish sacred forests around the planet where well run spiritual retreats are offered to religious devotees can sanctify and safeguard more wild places and protect more wildlife than all the biodiversity reserves and entrepreneurial developments extant. This will foster an explosion of exceptionally low-impact pilgrimages that swamp conventional aesthetic and adventure tourism.

I urge members of IUCN, and all concerned conservationists, to accept and embrace this new force. A first step in this regard would be to lobby for expansion in the focus of CITES (Conventional on International Trade in Endangered Species of Wild Fauna and Flora). When the CITES "scripture" was written in 1973 there was little or no representation from the religious institutions. Crucial points of view regarding the value of wild fauna and flora can be added to the CITES Preamble by altering the first two sentences and inserting these underlined words:

- RECOGNIZING that wild fauna and flora in their many <u>wonderful</u> and varied forms are an irreplaceable part of the natural systems of the earth which must be protected for this <u>time</u> and the <u>future</u> to come

- CONSCIOUS of the ever-growing value of wild fauna and flora from aesthetic, cultural, economic, recreational, <u>religious</u>, <u>spiritual</u>, and scientific points of view.

In the first line, substitution of 'wonderful' for 'beautiful' is more than cosmetic. It signifies the deep spiritual power of the natural world, inclusive of, but not limited to, the aesthetic. This invites the vast public that values more than surface appearance and variety to join. The shift to protecting flora and fauna for 'this time and the future' expands to include respect for nature's intrinsic values, and not merely its worth to the generations of humanity.

In line two I added 'religious' to honor the views of countless peoples and societies that rely on the presence of wild flora and fauna in their rituals and rites. The religious dimension has sometimes been included in 'cultural', but it is better separated. Cultural is used for homogeneous small groups and societies. Our great global religions are trans-cultural institutions that, as the religious environmental movement demonstrates, have many needs to connect with natural creation. A penultimate need is the 'spiritual'. Again, this is not solely the domain of cultures, nor of religions. One must add the spiritual point of view to honor the needs of individuals, families, and small groups for deep communion with those intangible powers of natural creation that sustain all of humanity.

These changes are proposed seriously in hopes that they will be made part of a conscious effort to include religious and spiritual concerns in all the arenas where the CITES accords are at play. With moral leadership of the world religious community inspiring business endowment of right action, conservationists can set aside worst case scenarios and shift to a "save them all" strategy. The leaders of the new conservation movement will not settle for saving small pockets of those species now said to be closest to extinction, nor will it continue to look aside while bushmeat orphans die.

The new leaders will challenge the narrow species fixation itself, and ask that all animal communities under threat of destruction be protected with emergency effort, while work is done to endow massive and far reaching life assurance systems to safeguard primates and other endangered orders. With the strategic focus of wildlife conservation shifted from gazetting biological arks to protecting all the major elements of natural creation from the human flood, a far different set of missions, goals, and objectives will be pursued by a new kind of conservation professional. Imagine the tactics and talent required to conserve the Congo Basin ecosystem and the primate order for all time as elements of Eden that are spiritually sacrosanct and financially secure! Those who lead us into this endeavor will need many thousands of committed workers to pursue the worthy success.

4.4   Ecosocial Practice Achieves

Endowed by international business and inspired by global religion, the conservation activist of the future will spring from a marriage of ecology and applied social sciences. These ecosocial practitioners will become the third force—the institutes and action teams that design, build, and manage local, regional, and global conservation organizations and programs (Rose, 2001a). The traditional conservation community should find this third force more acceptable than business and religion. Anthropologists and ecologists have collaborated to study and help indigenous peoples in wild environments. But my experience is the opposite. Most conservation biologists admit their ignorance about business and recognize the power of religious institutions, but everyone seems to think they are experts in analyzing and effecting social change.

The idea of protecting enclaves of apes without helping human society is not feasible. What is needed is to study, assess, and promote *biosynergy*— the continual synergistic relationships among ecological and social forces, processes, and stakeholders to assure that both humanity and nature will thrive (Rose, 2001b).

The expertise required to produce effective biosynergy in places like equatorial Africa is diverse and scattered at best. Professionals competent in all fields are required, from community builders to law enforcers. But to recruit and organize the best of them to work together for African wildlife conservation is a Herculean effort. I have drafted program designs for multi-faceted, multi-disciplinary, multi-level, community based ecosynergy projects to control bushmeat commerce and develop sustainable alternatives with people in Cameroon. Response from specialists has been consistent— "too complex, too costly, too many disciplines, too much territory to cover." Advice is common—"start with a simple community based pilot study to stop poaching in a small wildlife reserve."

When I ask "what will a project in a small reserve tell us about commercial hunting in huge logging concessions feeding urban markets?" many reply "start where you can succeed." They overlook the fact that success in small isolated projects is short-lived (Western, *et al.*, 1994). When I ask "how can a forest community stop poaching without outside support?" they say "governments and timber companies must enforce the law." They fail to explain how exploiters and politicians will be taught to infuse conservation values, develop ecosocial change projects, and govern and monitor huge concessions. Reaction from systems-oriented professionals is better: most recognize the value of large multivariate programs, but few are enthusiastic about joining interdisciplinary teams.

The promotion of biosynergy will become the mission of the new conservation movement. But to pursue that mission, methods for achieving synergy among teams of ecosocial professionals and representatives of stakeholder communities must be invented and installed. The barriers and prejudices that keep us apart and in conflict must be overcome first. If physical and cultural anthropologists are still at odds, how much more effort will be needed to unite sociologists, biologists, theologists, entrepreneurs, and economists?

There is no alternative. The old approach of basic science that tests uni-factored theory and method in controlled settings will not work. The free enterprise model with wildlife and biodiversity focused NGOs competing for limited market share has failed. The social movement that is engulfing conservation calls for international support by business and religion of regional and global change programs that will maximize the salvation of humanity and nature. To respond to this call, scores of professionals are needed with the courage and the will to collaborate with strange bedfellows in places where exploitation, migration, and conflagration are destroying people, wildlife, and environment.

5.   CONCLUSION: CONSERVATION MUST SERVE AND SYNERGIZE HUMANITY AND NATURE

Fast and durable success will come to innovative conservationists who work directly with the people involved in expanding human commerce, including poachers and traders, suppliers and producers, exploiters and consumers, leaders and rulers. These proactive partnerships will invent socially and ecologically synergistic programs to satisfy the human needs that now drive the commercial extraction and consumption of fauna and flora in Africa. Cadres of devoted ecosocial practitioners, inspired and endowed by religion and business, will take over center stage from the lone field biologists and anthropologists who have served as long suffering crusaders for wildlife. Media will expand beyond romantic images of scientist saints rescuing individual apes and will celebrate the entrepreneurs, educators, and innovators who help local and indigenous people to improve the quality of life by returning to a reverential and synergistic relationship with the environment.

The task of living in wild places to track gorillas and chimpanzees will take on huge added responsibility as synergistic conservation proliferates. Teams of professionals and community leaders will collaborate to convert poachers to protectors, monitor forest product and service sustainability, and implement ecosocial improvement projects. The study of nonhuman biology and behavior will be one of the forest services sustained in the long term by practical interventions to transform human morality, instill conservation values, and effect ecosocial accountability.   Some fallen idols and abandoned adventures will be mourned, but as time passes the sense of loss will be supplanted by the satisfaction that will come from saving and enriching the lives of more African primates than we can ever know.

This satisfaction will accrue to a general public in Africa and around the world that has claimed its kinship with nonhuman primates through personal interaction and supports the social movement to save wildlife and nature as our moral obligation and spiritual need.   Everyone will know that a perpetually rich and thriving African rain forest with its apes and other ancestors alive and well is worth far more now and in the future than bundles of wood and bushmeat.   Beyond the oxygen and medicine that the forests produce, and the lush beauty and mystery they provide, they give us profound insight into our identity.   After all, hominids came out of Africa.

The conservation *zeitgeist* of the 21st Century will explode into a humane and moral social movement that will be implemented by competent ecosocial practitioners and guided by five strategic imperatives.

1. Social and moral leaders will promote humanity's profound obligation to conserve wildlife and to restore the natural world.
2. Political and economic authority will place conservation on a par with human rights and welfare.
3. Conservationists will shift from measuring biodiversity to ensuring the biosynergy of humanity and nature.
4. Demand for religious and spiritual values of nature will overtake utilitarian exploitation and assure sustainable development.
5. All wildlife habitats will be considered sacrosanct, and all human intrusion and involvement will be managed in a moral, businesslike, and synergistic way for the global good.

The success of this great new social movement, this global life alliance, will do more than save wildlife and wilderness. It will safeguard the world ecology, restore biosynergy, and reinspire the natural spirit of humanity itself. As founders of the movement, conservationists must work together with a wealth of colleagues and fellow travelers, always in reverence, to celebrate the fulfillment of human origins and destiny in the vast and wonderful Creation that unfolds and evolves on this remarkable planet.

REFERENCES

Alexander, A., 1998, History of the Christian Environmental Council, talk to *Inter-Religious Council,* Mexico, April.

Allen, C. and Bekoff, M., 1997, *Species of Mind: The Philosophy and Biology of Cognitive Ethology,* Boston: MIT.

Ammann, K., 1993-4, Orphans of the forest (Parts I and II). SWARA, Nov.-Dec.: 16-19 and Jan.-Feb.: 13-14.

Ammann, K., 1994, The bushmeat babies, *BBC Wildlife,* Oct.: 16-24.

Ammann, K., 1996a, Primates in peril, *Outdoor Photographer,* February.

Ammann, K., 1996b, Timber and bushmeat industries are linked throughout west/central Africa. Talk at: *Seminaire sur l'impact de l'exploitation forestiere sur la faune sauvage,* Bertoua, Cameroon, April.

Ammann, K., 1996c, Halting the bushmeat trade: Saving the great apes. Talk at: *World Congress for Animals,* Wash. DC, June.

Ammann, K., 1996d, Conservation in central Africa: Time for a more business-like approach, *African Primates,* IUCN/SSC, 2(1): 30-31.

Ammann, K., 1997, *Gorillas,* Hong Kong: Insight Topics, Apa Publications (HK) Ltd.

Ammann, K., 1998a, The conservation status of the bonobo in the one million hectare Siforzal/Danzer logging concession in Central D.R.Congo, Bushmeat Project Website: http://bushmeat.net.

Ammann, K., 1998b, Personal correspondence and communication; 1988 to present, Hermosa Beach, CA

Ammann, K. and Pearce, J., 1995, *Slaughter of the Apes: How the Tropical Timber Industry is Devouring Africa's Great Apes,* London: World Society for the Protection of Animals.

Cartmill, M., 1993, *A View to a Death in the Morning: Hunting and Nature Through History,* Cambridge: Harvard University Press.

Cavalieri, P. and Singer, P., 1993, *The Great Ape Project: Equality Beyond Humanity,* New York: St. Martin's Press.

Cheney, D.L. and Seyfarth, R.M., 1992, *How Monkeys See the World: Inside the Mind of Another Species,* Chicago: University of Chicago Press.

Dupain, J. and van Elsacker, L., this volume.

Eltringham, S.K., 1984, *Wildlife Resources and Economic Development,* New York: John Wiley and Sons.

Fouts, R. and Mills, S.T., 1997, *Next of Kin: What Chimps Have Taught Me About Who We Are,* New York: William-Morrow.

Goodall, J., 1998, *The African Bushmeat Trade: A Recipe for Extinction,* Press Report, Ape Alliance, London.

Hawley, J.A., 1993, *Reawakening the Spirit in Work,* San Francisco: Barrett-Koehler.

Hennessey, A.B., 1995, *A Study of the Meat Trade in Ouesso, Republic of Congo,* Brazzavile: GTZ.

Incha Productions/ZSE-TV, 1996, *Twilight of the Apes,* ZSE-TV, Johannesburg (Video: 25 min.).

Juste, J., Fa, J.E., Del Val, J.P., and Castroviejo, J., 1995, Market dynamics of bushmeat species in Equatorial Guinea, *J. of Applied Ecology* 32: 454-67.

Kano, T. and Asato, R., 1994, Hunting pressure on chimpanzees and gorillas in the Motaba River area, northeastern Congo, *African Study Monographs* 15(3): 143-162.

Kellert, S.R., 1993, The biological basis for human values of nature. Pp. 42-47 in: (Eds. S.R. Kellert and E.O. Wilson), *The Biophilia Hypothesis,* Washington D.C.: Island Press.

Kellert, S.R., 1996, *The Value of Life: Biological Diversity and Human Society*, Washington D.C.: Island Press.

Kellert, S.R. and Wilson, E.O., 1993, *The Biophilia Hypothesis*, Washington D.C.: Island Press.

King, B.J., 1995, *The Information Continuum*, Santa Fe: SAR Press.

Maslow, A., 1993, *The Farther Reaches of Human Nature*, New York: Penquin-Arkana.

McRae, M. and Ammann, K., 1997, Road kill in Cameroon, *Natural History Magazine*, (106)1: 36-47, 74-75.

Mittermeier, R.A., 1987, Effects of hunting on rain forest primates. Pp. 109-146 in: *Primate Conservation in the Tropical Rain Forest*, New York: Alan R. Liss.

Mordi, R., 1991, *Attitudes Toward Wildlife in Botswana*, New York: Garland.

Oates, J.F., 1996a, Habitat alteration, hunting, and the conservation of folivorous primates in African forests, *Australian J. of Ecology* 21: 1-9.

Oates, J.F., 1996b, *African Primates: Status Survey and Action Plan* (Revised), IUCN, Gland.

Owens, D. and Owens, M., 1992, *The Eye of the Elephant: Epic Adventure in the African Wilderness*, Boston: Houghton Mifflin.

Patterson, F. and Linden, E., 1981, *The Education of Koko*, New York: Holt, Rinehart, and Winston.

Rogers, C.R., 1961, *On Becoming a Person*, New York: Houghton-Miflin.

Rose, A.L., 1994, Description and analysis of profound interspecies events. *Proceedings of XVth Congress of International Primatological Society*, Bali, Indonesia.

Rose, A.L., 1996a, Orangutans, science and collective reality. Pp. 29-40 in: (Eds. R. Nadler, B.M.F. Galdikas, L. Sheeran, and N. Rosen), *The Neglected Ape*, New York: Plenum Press.

Rose, A.L., 1996b, Commercial exploitation of great ape bushmeat. Pp. 18-20 in: (Eds. R. Ngoufo, J. Pearce, B. Yadji, D. Guele, and L. Lima), *Rapport du seminaire sur l'impact de l'exploitation forestiere sur la faune sauvage*, Bertoua: Cameroon MINEF and WSPA.

Rose, A.L., 1996c, The African forest bushmeat crisis: Report to ASP. *African Primates*, IUCN/SSC, 2(1): 32-34.

Rose, A.L., 1996d, The bushmeat crisis is conservation's first priority. Talk at: IUCN Primate Conservation Roundtable Discussion on an Action Agenda, at *XVIth Congress of IPS/ASP*, Madison, August.

Rose, A.L., 1996e, The African great ape bushmeat crisis, *Pan Africa News* 3(2): 1-6.

Rose, A.L., 1997, The African primate bushmeat crisis: Action workshop, *Annual Meeting of the American Society of Primatologists*, San Diego, June.

Rose, A.L., 1998a, Finding paradise in a hunting camp: Turning poachers to protectors, *J. of the Southwestern Anthropo. Assoc.* 38(3): 4-11.

Rose, A.L., 1998b, On tortoises, monkeys, and men. In: (Ed. M. Tobias), *Kinship with the Animals*, Hillsboro, OR: Beyond Words.

Rose, A.L., 1998c, On the road with a gorilla hunter, *Gorilla: Journal of the Gorilla Foundation* (21)1: 2-6.

Rose, A.L., 2000, Report from the field: Affecting empathy for apes, Bushmeat Project Website: http://bushmeat.net.

Rose, A.L., 2001a, Social change and social values in mitigating bushmeat commerce. In: (Eds. Bakarr, da Fonseca, Mittermeir, Rylands, and Painemilla), *Hunting and Bushmeat Utilization in the African Rain Forest,* Washington: Conservation International.

Rose, A.L., 2001b, The bushmeat crisis: Strategies, solutions, and social change capacity, Invited talk at 18th Congress of the International Primatological Society, Adelaide, Australia.

Rose, A.L., 2001c, Conservation must pursue human-nature biosynergy in the era of social chaos and bushmeat commerce. In: (Eds. A. Fuentes and L. Wolfe), *Conservation Implications of Human and Nonhuman Primate Interconnections*, Cambridge: Cambridge University Press.

Rose, A.L. and Ammann, K., 1996, The African great ape bushmeat crisis, Talk and workshop at *XVIth Congress of International Primatological Society / American Society of Primatologists*, Madison, August.

Rose, A.L. and Auw, A., 1974, *Growing Up Human*, New York: Harper and Row.

Rose-Erejon, J., 1998, Personal communication, Mexico City, May.

Savage-Rumbaugh, S. and Lewin, R., 1996, *Kanzi: The Ape at the Brink of the Human Mind*, New York: John Wiley & Sons.

Singer, P., and Cavalieri, P., this volume.

Splaney, L., 1998, Hunting is greater threat to primates than destruction of habitats, *New Scientist*, March, 18-19.

Steel, E.A., 1994, *Study of the Value and Volume of Bushmeat Commerce in Gabon*, Libreville: WWF and Gabon Ministry of Forests and Environment.

Stromayer, K. and Ekobo, A., 1991, *Biological surveys of southwest Cameroon*, Wildlife Conservation International.

Thompson, J.A., this volume.

de Waal, F., 1990, *Chimpanzee Politics*, Boston: Harvard Press.

de Waal, F., 1996, *Good Natured: Origins of Right and Wrong in Humans and Other Animals*, Boston: Harvard Press.

Western, D., Wright, M.R., and Strum, S.C. (Eds.), 1994, *Natural Connections: Perspectives in Community-Based Conservation*, Washington D.C.: Island Press.

Wilkie, D.S., Sidle, J.G., and Boundzanga, G.C., 1992, Mechanized logging, market hunting, and a bank loan in Congo, *Conservation Biology* 6(4): 570-580.

Wilson, E.O., 1984, *Biophilia: The Human Bond With Other Species*, Cambridge: Harvard University Press.

Wrangham, R. and Peterson, D., 1996, *Demonic Males: Apes and the Origins of Human Violence*, New York: Houghton-Mifflin.

Chapter 19

STATUS OF THE PROPOSED LOMAKO FOREST BONOBO RESERVE: A CASE STUDY OF THE BUSHMEAT TRADE

J. Dupain and L. Van Elsacker
Royal Zoological Society of Antwerp, Kon. Astridplein 26, B-2018 Antwerp, Belgium

1.   INTRODUCTION

The bonobo (*Pan paniscus*) is endemic to the Democratic Republic of the Congo (formerly Zaire). Its geographic range is limited to the left bank of the Congo River (Figure 55). The eastern boundary is thought to be the Lomami or Lualaba River; to the south, bonobo distribution is limited by the Kasai-Sankuru Rivers (Coolidge, 1933; Kano and Furuichi, 1984; Kortlandt, 1995). However, with the exception of some areas, the precise location of bonobos in this range is unknown. This lack of knowledge accounts for the large variation in overall bonobo population size estimates (Dupain and Van Elsacker, this volume). Recent publications agree on a figure of around 15,000 to 20,000 individuals, as proposed by the Bonobo/Pygmy Chimpanzee Protection Fund (Japan, 1992). All of these figures are based on the single extensive survey made by Dr. Kano in 1984 (Dupain and Van Elsacker, this volume). More accurate density information is currently available for some established study sites (Hashimoto and Furuichi, this volume; Thompson, this volume).

An important bonobo population is known to live in the Lomako Forest (Equateur, Dem. Rep. Congo (Figure 55)). Ever since the first bonobo studies in this forest, emphasis has been placed on the optimal conditions for making this forest a reserve (Badrian and Badrian, 1977, 1978; Susman *et al.*, 1981). At that time, however, the area was part of a

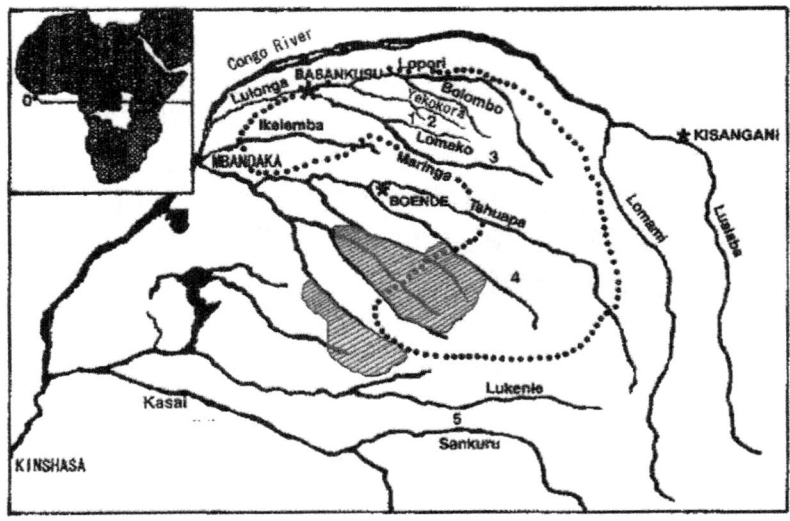

Figure 55. An important bonobo population is known to live in the Lomako Forest.

Siforco logging concession (Danzer Furnierwerke GMBH & Co). In 1987, the logging company abandoned the area and even offered their wharf along the Maringa river to World Wildlife Fund (WWF)-Germany (Susman, 1989). In 1990, WWF-International submitted a proposal to the former Institute Zairois pour la Conservation de la Nature to create a reserve of 3,800 km² (Figure 56). In May 1991, the proposal reached Ministry level, but due to political turmoil, it was never approved. The proposed reserve was thought to be undisturbed, suitable habitat for bonobos, and apart from the presence of researchers, without permanent human inhabitants (Thompson-Handler, *et al.,* 1995). In 1995, Thompson-Handler and her colleagues again stressed the urgent need for the creation of the Lomako Reserve. However, prior to any further progress more surveys of the region are urgently needed.

Within the proposed reserve, three bonobo research projects have been established (Figure 56): The Lomako Forest Pygmy Chimpanzee Project (LFPCP) (established by R. Susman, Stony Brook University, e.g. Susman, 1984) and the Isamondje study site (started in 1990 by G. Hohmann, Max Planck Institute, e.g. Fruth and Hohmann, 1993). Approximately 20 km from these two sites, the Iyema-Lomako (IYEMA) site was established as a base for the Bonobo *in Situ* Project (1995, Royal Zoological Society of Antwerp, e.g. Dupain, *et al.,* 1996a, b). All research in the Lomako Forest focused on the socioecology of the bonobo; little work was done on conservation issues.

In this chapter, we summarize the importance of the proposed Lomako Reserve, and focus on the relationship of the bushmeat trade to other conservation issues in this area. We report on the influence of logging on hunting pressure, describe the situation of the local people, and provide preliminary results of a bushmeat market study conducted near the Lomako Forest.

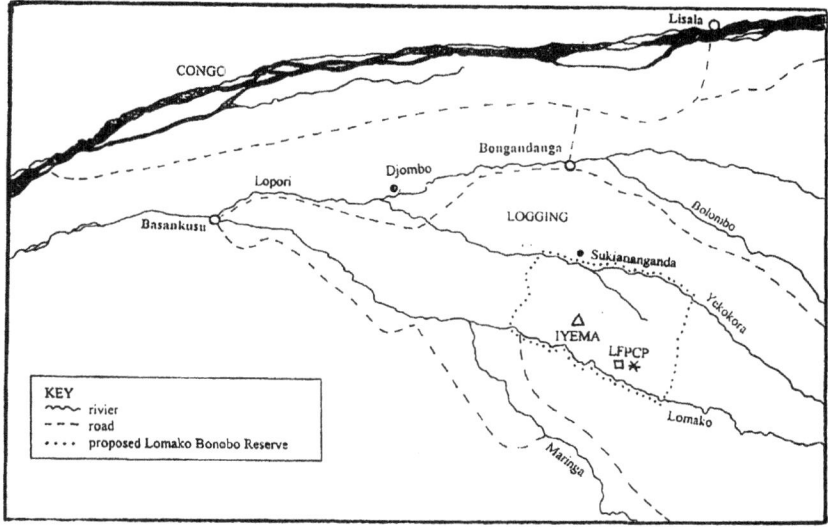

Figure 56. The proposed Lomako Forest Reserve. Multiple sites in the area have become trade centers for bushmeat.

2.　　THE IMPORTANCE OF THE LOMAKO FOREST

The Lomako Forest was identified in colonial times as an important area for nature conservation. In 1954, Father G. Hulstaert discussed the Lomako Forest in a letter to the President of the Institut des Parcs Nationaux de Congo Belge as a suitable area for the creation of a reserve. The Belgian government opted for the alternative proposition, that of creating the Salonga National Park. The features below highlight Lomako's suitability as a reserve.

1. Permanent habitation has been forbidden in this forest block since the 1920s, first by the colonial government and later by the Zairean government. The local population, the Mongo ethnic group, resides in officially sanctioned villages north of the Yekokora River or south of the Lomako River. The Mongo people are allowed to establish temporary fishing or hunting camps in the forest, but they are only permitted to stay there for a few weeks each year. According to the last *Action Plan for Pan paniscus* (Thompson-Handler, *et al.,* 1995), no humans inhabit the forest north of the existing study sites. The proposed reserve was therefore believed to be undisturbed terrain.

2. The American LFPCP along the Lomako River affords some protection to bonobos through the researchers' presence at the site. Prior to this long-term project, the Mongo, who hunted in this forest, apparently practiced a taboo against eating bonobos.

3. In addition to *Pan paniscus*, several other rare and endangered species such as the Congo peafowl (*Afropavo congensis*, Dupain *et al.,* 1996c), chevrotain (*Hyemoschus aquaticus),* and the African golden cat (*Profelis aurata*) are found in this area. Elephant and bongo are still present in the eastern part of the proposed reserve (J.D. pers. obs.).

4. Due to the difficulties in exporting logs along the Lomako or Yekokora Rivers, it is highly unlikely that any logging companies will be interested in exploiting this region in the near future (Siforco, pers. comm.; Sokinex, pers. comm.; Scibois, pers. comm.)

The convergence of the above elements appeared to indicate that the Lomako Forest was the ideal area for the creation of a reserve for bonobos and other fauna. Unfortunately, our data do not entirely validate this view.

3.   LOGGING'S INFLUENCE ON HUNTING AT LOMAKO

In 1995, the first author conducted a 45-day survey along the Yekokora River (the northern boundary of the proposed reserve). We found this region of the forest to be inhabited (Dupain, *et al.,* 2000). We met small groups of hunters, most of whom were either ethnic Ngombe from the north or from Basankusu, or people belonging to the religious sect of the Kitiwalists, living in the southeastern part of the proposed reserve. While the latter group has resided in this area since the mid-1960s, the Ngombe hunters were quite recently attracted to the area by the widespread stories of high animal densities in the Lomako Forest. We gathered clear evidence that the bushmeat they hunted was sold to the lumber company workers based north of the Yekokora River. A market was held on the border of the Yekokora River (Figures 56 and 57). There the hunters met lumber workers and their families and exchanged bushmeat for clothes, medicines, cartridges, and other goods. We did not at that time conduct a thorough investigation. However, in January 1998, Karl Ammann organized an expedition to the same region to gather more precise data. The main goal was to get an idea of the importance of bushmeat hunting in this area, with particular reference to the influence of the nearby lumber company. Team members (including J.D.) stayed for about one week near the marketplace. We visited the small settlements in the surrounding regions and compiled questionnaires on hunting practices and the organization of the bushmeat trade. After visiting the market, we searched on the Lopori River for boats transporting wood, and eventually bushmeat, to Kinshasa. Finally, we flew in a small plane over the logging concession to assess the visibility of logging activities from the air.

Figure 57. Lumber companies facilitate consumers and traders of bushmeat.
© J. Dupain (Bonobo-*in-Situ*, RZSA)

The main findings of this expedition are as follows (see also Ammann, 1998):
- Bonobos are still present in the area, but their density seems to be decreasing, while the area covered for hunting is increasing. We

accounted for the limited access to bonobo meat (seven fresh carcasses in one week) by the decreasing density of bonobos and by the scarcity of cartridges. The latter was a temporary problem attributed to the ongoing war in Congo-Brazzaville that caused the factory manufacturing MACC-cartridges in Pointe Noire to close. This is no longer true today. Moreover, cartridges are mainly provided by the lumber company, which also provides guns. During our stay, one of our local collaborators witnessed the arrival of a lumber company plane carrying a fresh supply of cartridges.

- The lumber company facilitates consumers and traders of bushmeat and provides them with a means of transport for this trade. Consumers and traders are the company's employees and their families.
- The lumber company attracts "outside" hunters. These are enterprising men, mainly belonging to the Ngombe ethnic group.
- Logging activity increases hunting pressure on the local fauna, not only in the logging concession, but also deep in uninhabited and "protected" (since no logging activities are planned) forest.

A commonly cited solution to these problems is the provision of meat to the lumber company employees (which is not done at all) and the prohibition of the transport of smoked bushmeat out of logging concession areas on company boats. Halting the import of cartridges may reduce the impact on bonobo and other primate populations in general, as these animals are usually hunted with guns. Snares are mainly used to trap ungulates and rodents (Delvingt, 1997).

4.   OTHER EXPLOITATION OF THE LOMAKO REGION

Even if logging activity was prohibited in the Lomako and surrounding region, there are no guarantees that commercial, unsustainable hunting would also stop. Defaunation of the northern part of the proposed reserve appears to be mainly the result of the diverse effects of timber trade, which attracts immigrant commercial hunters. The southern part suffers from exploitation of forest products as alternatives to the declining coffee, rubber, and palm oil businesses.

Traditional laws state that different sections of the forest block north of the Lomako River belong to the various villages situated south of the river. According to governmental law, these people may exploit the forest and may stay there for a few weeks, but permanent settlement is forbidden. As was noted in the last *Action Plan for Pan paniscus* (Thompson-Handler, *et al.*,

1995), the implementation of these laws is assumed. However, we calculated a permanent human density of 0.4 individuals/km² in about 250 km² of this "uninhabited" region (Dupain, unpublished information), and evidence indicates that the resident human population in the proposed Lomako Reserve is growing. Our interviews with the local people made it clear that this situation reflects what is happening throughout the 3,800 km² Lomako region. Furthermore, the situation has been made official by local authorities, which since 1996 have formally recognized most permanent settlements in the forest.

After some time spent talking with the local people, it was apparent that most of them preferred their former way of life in their natal villages where many of them had coffee, maize, or cocoa plantations. However, due to the deteriorating infrastructure of the country, businessmen from Kinshasa rarely or never visit these villages, and the villagers have no idea when boats will arrive along the Maringa River. Additionally, the forest fauna adjacent to their homes has been all but exterminated. In short, although most local people would prefer to stay in their natal villages, living close to their relatives with easy access to schools and other important services and resources, increasing numbers of them survive by hunting in the more distant forests. Thus, subsistence hunting escalates into commercial hunting, which enables the local people to trade bushmeat for clothes, medicines, soap, salt, and other necessities.

The phenomenon of remigration to the forest continues to put pressure on wildlife. In order to adress this problem, we plan to establish an agricultural program around the existing plantations. This program is at the request of, and in collaboration with, the local people and is supported by the Bonobo Protection Fund. We will serve as facilitators between businesspeople in Kinshasa and the villagers in the forest. In this way, the local people will be able to return to a preferred lifestyle in their natal villages, which should slow down migration into the forest. The program aims to reduce the need for commercial hunting activities. Simultaneously, we hope to demonstrate the indirect economic benefit of bonobo conservation. Several village heads in the surrounding areas contacted us in 1998 asking if there would still be a chance of attracting researchers to "their" part of the forest if bonobo hunting was stopped.

5. BUSHMEAT MARKETING

Although we might succeed in reducing commercial hunting activities in the proposed Lomako Reserve, the demand for meat in large cities will probably grow. All bushmeat hunted along the Lopori, Yekokora, Maringa, and Lomako Rivers is, if not consumed in the forest, mainly transported into Basankusu. This city has become a major market for bushmeat en route to Kinshasa. In order to acquire information on the hunting pressure in the forests adjacent to these rivers, we have started to monitor the available meat at the two markets in Basankusu (Figure 58). Table 20 shows some preliminary results. In one month (February 1998), we counted a total of 749 carcasses. Bonobo carcasses represented less than 1% of the total number. About 53% of the animals were ungulates and 31% were primates. These percentages are similar to what was found in other market studies (see for review Bowen-Jones, 1998). We also encountered three infant bonobos, two of them on a boat heading for Kinshasa (Figure 59). People we met during our expedition in January 1998 complained of a lack of cartridges. Taking this into account, it is likely that the number of bonobo carcasses counted in Basankusu may have been lower than during periods when there was no shortage of cartridges.

We used questionnaires to determine the origins of 500 animal carcasses (Table 21, Figure 56). The four bonobo carcasses counted all originated from the Lomako region. In fact, 36% (n=178) of all carcasses came from the Lomako region; only 3% (n=17) were hunted near the Yekokora River. The lower number from Yekokora reflects the fact that most bushmeat there goes directly to lumber company employees (Figure 60). Future questionnaires should explore in detail the exact hunting location, the identity of the hunter, the means of hunting, and other details, but our preliminary results highlight the importance of the Lomako region as a source of animal protein for people living in large cities such as Basankusu. Currently, we are developing questionnaires on food preferences, profit margins at the market, and related variables.

Figure 58. In Basankusu, questionnaires on food preferences and profit margins at the market are developed. © J. Dupain (Bonobo-*in-Situ*, RZSA)

J. Dupain and L. Van Elsacker

Table 20. Preliminary findings of the bushmeat market study at Basankusu (one month). Total number of carcasses counted = 749.

	Genus / Species (if known)	# (% of total)
Order Artiodactyla	*Cephalopus callypigus*	128
	Cephalopus nigrifrons	46
	Cephalopus monticola	37
	Cephalopus dorsalis	26
	Cephalopus sylvicultor	22
	Potamochoerus porcus	98
	Hyemoschus aquaticus	21
	Tragelaphus spekei	18
	Total	**396** (52.87%)
Order Primates	*Piliocolobus tholloni*	22
	Colobus angloensis	58
	Cercopithecus ascanius	83
	Cercocebus aterrimus	43
	Cercopithecus mona	10
	Cercopithecus neglectus	7
	Cercopithecus nigroviridis	2
	Pan paniscus	4
	Total	**229** (30.57%)
Order Rodentia	*Atherurus africanus*	54
	Sciuridae	10
	Total	**64** (8.54%)
Class Reptilia	*Crocodylidae*	51
	Testudinidae	4
	Varanidae	1
	Total	**56** (7.48%)
Order Pholidota	*Manis tetradactyla*	**3** (0.40%)
Class Aves	Unidentified	**1** (0.13%)

Figure 59. In Basankusu, a bonobo orphan as by-product of the bushmeat trade.
© J. Dupain (Bonobo-*in-Situ*, RZSA)

Figure 60. In 1995, smoked meat was readily available in the hunting camps along the Yekokora River. © J. Dupain (Bonobo-*in-Situ*, RZSA)

Table 21. Origin of carcasses counted at the Basankusu market (n=495)

Location	Number of Carcasses
Along the Lopori River	88
Near Bongandanga	40
Along the Yekokora River	17
Along the Maringa River	38
Surroundings of Basankusu	134
Along the Lomako River	178
Total	**495**

6.   CONCLUSIONS

The most recent data on the status of the proposed Lomako Forest bonobo reserve are summarized here:

1. The Lomako Forest is an important area that deserves special conservation attention. It contains a significant population of bonobos, and it harbors other endangered and rare species. However, in a way that is not readily apparent to the outside world, it is subject to increasing human hunting pressure.

2. Hunting pressure is strongest in the northern part of the forest. Much of the bushmeat is sold to lumber company employees and their families, who also transport this meat to major cities. Importation of cartridges on company boats increases hunting pressure on larger species such as the bonobo. Providing workers with alternative foods and prohibiting the transport of meat and cartridges on company boats could greatly reduce hunting pressures due to the logging activities.

3. In the southern sector of the forest (and into the northern sector when we deal with the local population *sensu strictu*), hunting pressure is growing due to forced migration into the forest, caused by the difficulty of subsistence in the local people's natal villages. We hope to reverse this trend by promoting their former agricultural activities. This is occuring in collaboration with, and by request of, the local population.

4. Basankusu is an important urban center of bushmeat trade with regard to the fauna hunted in the Lomako Forest. Many people rely on the Lomako Forest, not only for meat consumption, but also as a source of income. Further study through market censusing and questionnaires is necessary to strengthen justifications for the creation of the Lomako Reserve.

7.   POSTSCRIPT (SEPTEMBER 2000)

At the time of departure from our research site in November 1998, Siforco had halted logging activities, and logging in the Lomako area has not been resumed. However, other timber companies have already begun to express an interest in setting up logging concessions in and around the proposed Reserve. Although the Lomako Forest officially belongs to the area controlled by President Kabila, it lies close to the area controlled by the Mouvement pour la Liberation du Congo. Since the beginning of 2000, contact with our local collaborators in the forest has been virtually impossible. Any suggestion regarding the status of the bonobo at this present time would be no more than a guess. We continue, however, to promote the establishment of the Lomako Reserve. Through these and other procedures, we aim to enhance the survival prospects of the local human populations and the flora and fauna of the Lomako Forest.

ACKNOWLEDGMENTS

We thank Gary Shapiro and Biruté Galdikas for their invitation to participate in the Great Apes of the World Conference and to contribute to this volume. We thank the Ministère de l'Environnement, Protection de la Nature et Tourisme and the Ministère de l'Enseignement Supérieur, Recherche Scientifique et Technologies, Kinshasa, République Démocratique du Congo, for providing authorizations and mission orders. All fieldwork was made possible through the financial support of the KBC. Further financial support was provided by the Wildlife Conservation Society, the LSB Leakey Foundation, the Fonds Leopold 111 pour l'Exploration et la Conservation de la Nature, and the Bonobo EEP members. Logistical support was provided by the Belgian Embassy and Cooperation in Kinshasa (Dem. Rep. Congo), CDI-Bwamanda, Claudine Minesi, Pierre Verhaeghe (AAC), Paul DePetter and J.Cl. Hoolans (Nocafex), Father Paul (Procure Saint-Anne, Kinshasa), and the missionaries of Mill Hill (Basankusu). We thank Karl Ammann and Reinhard Behrend for their invitation to join the January 1998 expedition. The research would not have been possible without the financial and other help of the Royal Zoological Society of Antwerp and the continuous support of Dir. F.J. Daman and Prof. R.F. Verheyen. We thank Lourdes Trujillo for helping to prepare this manuscript and Hellen Attwater for the editing. The first author was generously supported by the Center of Excellence Research as a visiting research scholar at the Primate Research Institute (Kyoto University) during the writing of this chapter.

REFERENCES

Ammann, K., 1998, The conservation status of the bonobo in the one million hectare Siforal/Danzer logging concession in central D.R. Congo. Eletronic document, http://biosynergy.org/bushmeat/, accessed July 1998.

Badrian, A. and Badrian, N., 1977, Pygmy chimpanzees, *Oryx* 13: 463-468.

Badrian, A. and Badrian, N., 1978, Wild bonobos of Zaire, *Wildlife News* 13: 12-16.

Bonobo/Pygmy Chimpanzee Protection Fund (Japan), 1992, *A Plan for the Protection of Bonobos (Pygmy Chimpanzee) of the Upper Luo Region.*

Bowen-Jones, A., 1998, A review of the commercial bushmeat trade with emphasis on Central/West Africa and the great apes. Report for the Ape Alliance c/o Fauna & Flora International, Cambridge, UK.

Coolidge, H.J., 1933, *Pan paniscus* (pygmy chimpanzee) from south of the Congo River, *Amer. J. Phys. Anthropol.* 8(1): 1-57.

Delvingt, W., 1997, La chasse villageoise. Synthèse réginal des études réalisées durant la première phase du Programme ECOFAC au Cameroun, au Congo et en République Centrafricaine. *ECOFAC/AGRECO.* Brussels, Belgium.

Dupain, J. and Van Elsacker, L., this volume.

Dupain, J., Van Krunkelsven, E., Van Elsacker, L., and Verheyen, R.F., 1996a, Iyema: A new field site for bonobo (*Pan paniscus*) Research. In: *Abstracts of the 1° Congreso de la Asociación Primatológica Española, APE 96*, European Workshop on Primate Research. Madrid, Spain, October 16-18, 1996: 28.

Dupain, J., Van Krunkelsven, E., Van Elsacker, L., and Verheyen, R.F., 1996b, The bonobo (*Pan paniscus*): Victim of human adaptation. In: *Abstracts of the Third Benelux Congress of Zoology*. Namen, Belgium, November 8-9, 1996: 12.

Dupain, J., Van Krunkelsven, E., Van Elsacker, L., and Verheyen, R.F. 1996c. Observations of Congo Peafowl (*Afropavo congolensis*) at the Equateur Province – Zaire, *Ostrich* 67: 46-47.

Dupain, J., Van Krunkelsven, E., Van Elsacker, L., and Verheyen, R.F., 2000, Current status of the bonobo (*Pan paniscus*) in the proposed Lomako Reserve (Democratic Republic of Congo), *Biological Conservation* 94: 254-272.

Fruth, B. and Hohmann, G., 1993, Ecological and behavioral aspects of nest building in wild bonobos (*Pan paniscus*), *Ethology* 94: 113-126.

Hashimoto, C., and Furuichi, T., this volume.

Kano, T. and Furuichi, T., 1984, Distribution of pygmy chimpanzees (*Pan paniscus*) in the Central Zaire Basin, *Folia Primatol.* 43: 36-52.

Kortlandt, A., 1995, A survey of the geographical range, habitats and conservation of the pygmy chimpanzee, *Primate Conservation*, 16, 1995: 21-36.

Susman, R.L., 1984, *The Pygmy Chimpanzee: Evolutionary Biology and Behavior*, New York: Plenum.

Susman, R.L., 1989, Auf den spruen unserer urahen, *Holz Aktuell* 7: 63-69.

Susman, R.L., Badrian, A., Badrian, N., and Handler, N., 1981, Pygmy chimpanzees in peril, *Oryx* 16: 179-183.

Thompson, J., this volume.

Thompson-Handler, N., Malenky, R.K., and Reinartz, G.E., 1995, *Action Plan for Pan paniscus: Report on Free-ranging Populations and Proposals for their Preservation*. Milwaukee, Wisconsin: Zoological Society of Milwaukee County.

All Apes Great and Small Volume I: African Apes, Co-Edited by Galdikas, Erickson Briggs, Sheeran, Shapiro, & Goodall, Kluwer Academic/Plenum Publishers, 2001

Chapter 20

WHAT HAPPENED TO *GORILLA GORILLA UELLENSIS?*: A PRELIMINARY INVESTIGATION

Karl Ammann[1] and Nancy Briggs[2]
[1] *Photojournalist, Nairobi, Kenya, Africa*

[2] *California State University Long Beach, Professor of Communication Studies*

1.   INTRODUCTION

Recent research suggests that Ethiopian hominid fossils are dated to at least 5.4 million years ago, which is one million years older than any other known human ancestor. This date is close to the point when the human and chimpanzee lines diverged (but see Janke and Arnason, this volume). Chimpanzees and gorillas are characterized by some distinctive features that apparently evolved after they each diverged from the common ancestor. These two African apes also share some features in common, such as the opposable big toe, which were apparently retained in their evolutionary lines but were lost during the course of human evolution. However, as hominoids, all three species share the opposable thumb and an enlarged brain, features all three species apparently retained from their common ancestor. The chimpanzee's genetic code is 98.6% the same as that of humans, while the corresponding value for gorillas is 97.7%. The living hominoids share much in common; hence, scholars' preoccupation with early apes and their evolution is understandable.

Taxonomists often separate modern humans (*Homo sapiens*) and their ancestors into the hominid group. Considerable effort has been devoted to recovering and analyzing hominid remains, but attempts to trace the fossil records of the African apes have met with little success. Here we describe a preliminary investigation for evidence of living and/or fossil populations of African apes in the Democratic Republic of Congo (DRC).

Many of the most exciting primatological expeditions are motivated by the underlying need for basic data on little-known species. The mystery surrounding *Gorilla gorilla uellensis* has not been addressed in research on the currently recognized gorilla species and subspecies. For example, distribution maps usually show only accepted distribution ranges for the western and eastern gorilla populations. At the end of the last millennium, there were still significant questions surrounding the possible existence of previously unknown gorilla populations. Undoubtedly, several unexplored areas of the DRC, the second largest nation in Africa, remained. *G. g. uellensis* may have existed (and may still exist) in the more remote regions of the DRC. However, the book *Gorillas* (Ammann, 1997) did not include any of these possibilities in its gorilla distribution range map despite the author's wishes. The publishers and editors felt it was a question of "playing it safe" and not raising unanswerable or difficult issues. More substantive information was needed to resolve the question of the subspecies' existence.

2.   HISTORICAL BACKGROUND ON *G. g. uellensis*

In 1898, a Belgian Army officer named Le Marinel collected four gorilla skulls from the Bondo area. These skulls were later placed in the Trevuren Museum in Brussels. In 1927 the museum curator, Henri Schoutenden, classified the skulls into a new subspecies, *G. g. uellensis*. He based his analysis on anatomical differences (when compared to other gorilla specimens) and their unique origin, which was some 640 km from the edge of the nearest western or eastern gorilla populations. Two years later American primatologist Harold J. Coolidge rejected Schoutenden's subspecies designation. Coolidge argued that it was well known that no gorillas existed at Bondo, and that it was more likely that the skulls had been brought into the region from elsewhere. Since then, taxonomists and biogeographers rarely discuss the possible existence of the subspecies.

3.   THE SEARCH FOR *G. g. uellensis*

One of us (KA, referred to hearafter in first person) was a member of a research party to find more skulls of this elusive ape and to ask local hunters about the possible presence of any "strange" apes. We brought along books with illustrations of all of the great apes and their habitats. We soon found that even some of the locally famous hunters had difficulty differentiating pictures of gorillas, chimpanzees, and bonobos, thus calling into question their ability to make finer distinctions within a given ape species.

The Azande language has four different words for chimpanzee, which translated include "the ape that beats the tree" and "the ape that kills the lion". Further, the hunters differentiated between large and small chimpanzees, but there were no stories about gorillas. We expected to be regaled with stories of hunters' encounters with aggressive apes, but we had yet to meet a hunter who described being charged by an irate silverback gorilla. The unanimous opinion of the survey team members was that we should look further east, near Bili, where the forest opens up into savannah and where many more species are found.

Our research team traveled to Bili. While showing hunters ape illustrations, we also looked for old skulls similar to the ones Le Marinel had found–in his case, in an abandoned hut. Eventually, a hunter at Bambillo showed us several skull specimens that were clearly collected from the village garbage heap. One of these skulls had a pronounced sagittal and even an occipital crest, but it was clearly too small to be a male gorilla's skull. We were quite excited and took the skull back to Kinshasa. In Kinshasa we obtained government authorization to take it to Kenya to have some casts made before it would be returned to the DRC. However, once taxonomists had the opportunity to examine the casts, the verdict was clear: it was a chimpanzee skull, albeit an unusual one.

While at Bondo, one of us (KA) contacted mission employees, who agreed to continue the search for *G. g. uellensis*. I received occasional letters with tantalizing tales and requests for more funds to continue the research and field study. I sent a few checks to a mission account in France but soon decided to discontinue this practice in case the funds were being used for other purposes. Moreover, visiting that region of the DRC was no longer an option due to escalating civil unrest.

In mid-1998, I decided to try a new approach to this problem. I retained a Cameroon hunter, an expert gorilla tracker, whom I befriended during various trips to explore the bushmeat trade. I asked the tracker to travel east through the CAR to do some research in the forests around Bili. The tracker never got beyond a village called Badai, some 30 km from Bili at the edge of the savannah. The local villagers warned him against proceeding due to the poor security situations prevailing in most urban centers. He stayed in Badai for a week and accompanied local hunters on their outings. He came across an abandoned ground nest that he thought was made by a gorilla and sent me a picture of it. I was not impressed by what I saw, but it did give me a new approach to use when questioning the local hunters about the presence of gorillas: look for ground nests as well as the apes themselves.

During 1999, most of northern Congo was taken over by a Ugandan-backed rebel force, the Movement for the Liberation of Congo, headed by Jean Pierre Bemba, an ex-businessman from Kinshasa. Through various contacts, I was able to get in touch with Bemba and ask him about the possibility of spending some time in and around Bili to continue research. Bemba answered that this would not be difficult to arrange.

In October 1998, I headed out with a Congolese friend, Louis Genezele, who was born in the region and had lived there most of his life. We flew into Zemio (CAR) and then made their way overland to Bili, checking in with a Uganda army contingent at Ango and presenting their credentials. The objective of this trip was to check the logistics for a documentary film covering the search for this lost gorilla population. The broken-down log bridges and the lack of operational car ferries on the larger rivers soon led us to the conclusion that the only productive way to complete the film would be to fly into the area. When we got to Bili, we started work on an airstrip. The results were not very different from the 1996 trip, except we kept hearing about areas where hunters regularly found ground nests.

Once we had cleared an airstrip, I called in an aircraft to come and pick me up. I left Louis behind to set up campsites and sort out logistics for two camps in areas where the hunters reported seeing these ground nests. One was approximately 20 km from Badai in a very remote forest island and necessitated cutting about 15 km of road.

In early December 1999, I went back to pick up Louis and inspect the campsites. I spent five days at the campsite near Badai. During long walks through the now burned out savannah and forest, I was shown some old ground nests, and team members did come across two fresh-looking nests. I taught the trackers how to collect hair samples from the nests while wearing rubber gloves to avoid contamination.

Back in Kenya, I decided to ship the hair samples to the National Institute of Health, under their CITES import permit, to find out if any DNA could be extracted. I was certain that for hair a CITES export permit would be needed, but I was also aware that he would never get one in Kinshasa if he told officials them that the hair had come from a rebel-held area. I did get an export permit from the local MLC authorities. When the shipment reached John F. Kennedy Airport (New York), it was confiscated by the U.S. Fish and Wildlife Services on the grounds that there was no acceptable export permit in place. Pleading with the authorities did not help, and the hair was reportedly destroyed.

The documentary film shoot was hinging on the likelihood of actually finding gorillas in Badai, and in this context positive DNA results would go far toward reassuring the filmmakers that the gorillas existed and were a worthy topic for a film. In mid-January, the trackers sent a radio message saying that they had found two groupings of nests from which they collected seven samples of hair, one of them consisting of long "white and silver" hair. I returned to inspect the nests myself and retrieve the hair samples.

In coordinating flight plans, I contacted Jean Pierre Bemba to arrange flight clearance via the Ugandan military for a routing from Entebbe straight to Bili. This time Bemba was unable or unwilling to cooperate, so there was now a serious problem. I had intelligence reports indicating that the Kabila army might attempt an invasion from the north using small CAR airstrips along the border. Bemba asked me to come to his headquarters at Gbadolite to discuss things. I flew via Douala to Bangui and then chartered a flight to Mobay across the border in the CAR then crossing into the DRC by pirogue over the Ubangi River.

In meeting with the rebel chief, it became clear to me that Bemba had some real doubts about the existence of the gorillas I sought and my motivation for being in the DRC. Gold and diamonds are generally the reasons why a foreigner would want to trudge through uncharted forests. Fortunately, I brought along two books: one of his gorilla books and *Orangutan Odyssey* (Galdikas and Briggs, 1999). These books seemed to make the difference: Bemba saw me as a researcher. In the end, he had no problem with my attempts to proceed to Bili. I made it to Bangassou on the CAR side where one of our trackers met me with the seven hair samples.

Traveling out of Africa with the samples meant layovers in two more Central African countries besides the DRC from where I now had an export permit from Mr. Bemba but none from Bemba's adversaries in Kinshasa. In Kenya, I convinced the local CITES management authorities to investigate and eventually accept Bemba's permit. When the authorities checked with the secretariat in Geneva, the advice back was to classify the samples as hair to be identified, not necessarily as samples from a CITES-listed mammal. Alternatively, the samples could be classified as "seized" and then sent under the Kenyan management authority export permit to request analysis. The second option proved to be the more feasible one.

Several laboratories in Europe and the U.S. received the hair samples. A gorilla taxonomy expert also analyzed the hair sample morphologically and came to the conclusion that at least the gray-silver hair sample was that of a gorilla or a gorilla-like creature. However, when the DNA test results were finalized, they did not confirm the above findings. The results indicated that the samples came from chimpanzees. Two laboratories could not amplify the DNA (which is normally considered very easy when the sample is chimpanzee hair but is difficult to accomplish with gorilla hair samples, suggesting that these samples might be from a gorilla).

Either the taxonomist was wrong, or the DNA evidence was incorrect. Although DNA analysis is supposedly an exact science, we were concerned about possible sample contamination or mix-ups. There was no choice but to go back to the DRC with the objective of collecting fecal samples. Fecal samples do not come under CITES and generally yield better DNA with which to work.

This time Jean Pierre Bemba approved the flight plans into and out of Bili. After fuelling stops at Entebbe and Arua in Uganda and Zemio in the CAR, my plane landed at Bili. The next day, appropriately enough on Good Friday, I reached camp and with the trackers found the first fresh nest sites in the very center of the streambed some 6 km from camp. Luck was with the team—we also found a very fresh and large fecal sample on the rim of the nest and various foot and knuckle prints nearby.

The team now included another experienced hunter, who in his lifetime had lived in other regions including Adama and Bondo. He confirmed that nowhere else, except for some of the riverbeds in the forest island nearby, had he encountered these large ground nests. He added that nowhere else had he noticed the feeding of fresh Rafia palm shoots by chimpanzees, although it was clear that he was quite familiar with chimpanzee ecology and behavior. He showed us a chimpanzee stick fishing site of red ants in a nearby riverbed.

The shape of the feces showed the ring-like compartments that normally characterize gorilla excrement. The size and shape of the feet and knuckle prints, which I cast in plaster, seemed to support the existence of a gorilla or gorilla-like creature in the area.

I carefully photographed and filmed this evidence, packed the casts for transport, and got the fecal sample by motorbike to Bili where there was a kerosene-powered refrigerator. The coming days in the field were less productive, although I did find some old nests and asked the hunters to show the nest site where in mid-January they had found the original hair samples. The nests were still visible. They were made of the same type of wild ginger plant and were similarly situated in the bottom of a streambed. I was by now certain that the chimpanzees in the area also made ground nests. In addition, trackers found some isolated nests at higher and drier elevations and also found correspondingly aged tree nests with them, which was not the case with the nests in the streambeds.

When Louis came back from Bili (where he took the fecal samples), he had more news about the hair samples. Another laboratory had identified some of the black hair found by the trackers as that of a chimpanzee.

Despite these discouraging reports, Louis and I believed we finally had found conclusive evidence of the existence of a new African ape subspecies. We decided that in the light of the footprint casts, we would celebrate the existence of *G. g. uellensis.* Alternatively, if the DNA experts could not confirm our assessment and beliefs, we would then celebrate the discovery of a very distinctive chimpanzee.

4.   THE FATE OF *G. g. uellensis*

To help resolve the issue, I left a remotely triggered infrared camera in the area where the footprints were found. Whatever the DNA experts might conclude, I would not give up until I had an image of the creature in question. I also left still and video cameras with the trackers to see if, with a smaller party and less noise, they might get a chance to photograph or film the ape in question. The time for the expedition was running out: the rains had started in earnest and soon the savannah patches would again consist of 3 meter-high elephant grass and the streambeds would become impassable muddy valleys.

Back in Kenya, I spent a week obtaining the permit to ship the fecal material to the United States packaged in dry ice. Two of the samples went to German laboratories with much less red tape. With four laboratories currently examining the samples and with cameras on site, we should soon have a definitive answer to the question "What happened to *G. g. uellensis?*"

REFERENCES

Ammann, K., 1997, *Gorillas*. Singapore: Apa Publications.
Galdikas, B.M.F. and Briggs, N., 1999, *Orangutan Odyssey*, New York: Abrams Press.
Janke, A. and Arnason, U., this volume.

Chapter 21

APES, PERSONS, AND BIOETHICS

P. Singer[1] and P. Cavalieri[2]
[1] *Center for Human Values, Princeton University, Princeton, NJ 08544*

[2] *Etica & Animali, Milan, Italy*

1. INTRODUCTION

In the seventeenth century René Descartes argued for a fundamental separation between human and nonhuman animals, a separation reflecting the sharp distinction he drew on a more general level between thought and matter. While we have bodies and minds, animals only have bodies, and thus are locked into the inferior realm of matter. This absolute separation between humans and animals underlies the absolute dismissal, in Western thought, of animals from the sphere of moral concern. But Cartesianism is not the only doctrine in the history of Western moral philosophy that denies animals any intrinsic moral status. Kant's moral philosophy also makes a radical break between rational beings and the domain of mere things, which includes animals. Kant argues that because only humans have reason and can be autonomous, only humans are ends in themselves. All other things, including all nonhuman animals, are mere means to human ends. True, Kant finds an indirect argument against cruelty to animals based on the claim that those who are cruel to animals may end up being cruel to humans as well, but this counts for very little, given Kant's denial that the suffering of an animal is, in itself, a reason not to be cruel to it. As recently as 1992, the French philosopher Luc Ferry showed the continuing influence of Kant's perspective when he defended the idea of a sharp moral distinction between humans and other animals by asserting that only humans can be free, in the sense of being detached from the natural world (Ferry, 1992).

2.   WHO IS A PERSON?

Modern political and social philosophy takes intra-human equality for granted. This equality is normally expressed as equality between persons. Political philosophers seem to regard the concept of person as unproblematic, but this is not so in bioethics. Though the word "person" is, in current parlance, often used as if it meant the same thing as "human being", the terms are generally not viewed as equivalent in bioethical discussions. Most authors in fact use "person" to refer to a being possessing certain characteristics such as self-awareness and rationality; used in this way, the term is distinct from the biological sense of "human", meaning belonging to the species *Homo sapiens*. The relevant literature in bioethics is studded with discussions about whether the beings involved in various moral dilemmas are persons. Can fetuses or infants be considered persons? If so, at what point in their development? Are the severely intellectually disabled or the irreversibly comatose still persons?

What is it that makes these debates so heated? A distinction is often made between two uses of "person", the descriptive (also sometimes labeled as metaphysical) and the normative (or moral). On this view, to say of some being that she is a person in the descriptive sense is to convey some information about what the being is like, and this can amount to saying that she has particular characteristics; on the other hand, to use the term "person" in a normative way is to use it simply to ascribe moral properties, usually some rights or duties, and frequently the right to life to the being so denominated. It is with reference to this normative meaning that it is so important to decide, in ethical disputes, which of the beings involved in a dilemma are persons and which are not. The question is one of moral ranking because of the import we attach to the presence of personhood with respect to the wrongness of inflicting some specific harms, and in particular the wrongness of killing.

"Person" is a suggestive word. It comes from Latin in the form of *persona*, but it also has its corresponding term in Greek, *pròsopon*. It initially meant a mask worn by an actor in classical drama, and consequently came to refer to the character that the actor portrayed. During the Early Roman Empire, the word was first introduced into philosophical jargon by the Stoic philosopher Epictetus, who used it to mean the role one is called to play in life. Not only the very idea of a role, but also the reference to a task to be accomplished, point towards an interpretation of the concept of person in terms of a *subject of relations* (Trendelenburg, 1910).

This emphasis on relation is fundamental. Just as a role has more to do with the place a being occupies than with its nature, relation is a category that was seen, starting at least from Aristotle, as something added to, and thus accidentally and not essentially connected to, the substance of the thing. The manner in which the concept of "person" could thus float free of any definite metaphysical substratum made the concept particularly useful in early Christian theological controversies about the dogma of the Trinity, where the problem lay in expressing the relations between God and the Word (Christ), and between them and the Holy Ghost. After the settling of the dispute through the Council of Nicea, which stated that the Trinity is one substance and three persons, a counter-process took place, characterized by the attempt to remove from the notion of person the aspect of relation and by the insistence on its substantiality. Hence the famous definition of person given in the sixth century by the Latin philosopher Boethius: "Person is an individual substance of rational nature". But theology had already saved the life of the concept of person: its use in connection with God had prevented it from being used so that it could refer only to human nature, or from simply becoming another term for "human being". Although this identification continued to be made in the subsequent course of philosophical reflection, its arbitrariness is shown by the fact that after we know that a given being is a human being, it remains an open question whether that being is a person.

"Person" is thus a pivotal term. It is connected with two linked key notions of pre-modern ethics, the notion of role and the notion of relation. To attach great importance to the roles in which people interact, and to see morality as a set of orientations for establishing and maintaining the health of relationships, are features of traditional, small-scale societies that reappear in different ways and with a different emphasis in much of (and not only) our cultural history (Silberbauer, 1991). Homeric virtues can be identified only by identifying the basic roles of Homeric society and their specific requirements. Though Aristotle locates the ultimate human *telos* in metaphysical contemplation, in his functional concept of "man" a prominent part is played by the idea of "man" as a political animal, whose virtues are connected with a set of roles and with healthy relationships within the community or polis (MacIntyre, 1981). Classical natural law, on the other hand, sees humans as created to play a part in a divinely ordained community expressing God's glory, and morality as teaching what that part is. From this perspective, then, the notion of person is conservative. But there is a sense in which one can claim the contrary. "Person" does embody two ancient notions, but it does it in a modern way. The specific requirements of the role, as well as the specific objects of the relations, are not there. What is there is the abstract idea of role and relation, and this is a mark of modernity. Modern rational ethics emerges from a twofold process of abstraction: abstraction from any preconceived teleological or metaphysical framework, culminating in Henry Sidgwick's policy of keeping the nonethical commitments of moral philosophy to an absolute minimum; and abstraction in regard to the constituency of ethics, which in modern rational ethics is formed by generic individuals without distinction of sex, race, attitudes, or merits, moving in an anonymous social arena rather than standing at the center of a web of specific connections (Sidgwick, 1907). It is because it is particularly well suited to this context that the notion of "person" has taken on such importance in the current philosophical debate. "Person" refers to the idea of being a locus of relations and playing a role in the ongoing drama. This makes it typically evaluative in nature. Hence, to regard an entity as a person is to attribute a special kind of value to that entity. Moreover, since being a locus of relationships and playing a role is an accidental, rather than an essential, property of an entity, the range of application of the term can vary, thus providing a tool for moral reform.

It should now be clear why, when we initially developed the idea of what is now a book, *The Great Ape Project* (1993), we turned to the concept of person. What we aimed at was a first attempt at overcoming the species barrier. The book is meant to build up a collective case for granting to some nonhuman animals the same basic moral status that we humans currently enjoy. Just as the best candidates for such a step are, as we shall show in more detail shortly, the four species that are closest to us in the evolutionary tree (chimpanzees, bonobos, gorillas, and orangutans), so the best theoretical instrument is this traditional notion of a person. Though the Declaration on Great Apes, around which the book is organized, speaks in terms of admitting the other apes into the community of equals, it is around the concept of person that many of the arguments in the essays implicitly or explicitly revolve.

Granted, there are differences in the interpretation of the concept of person. To be effective, it must be given some more specific content, and its primary meaning needs to be articulated in more detail. Historically, this has been done in many ways. The connection with the capacity for reasoning that we have already seen in Boethius was taken up by Aquinas ("every individual of rational nature is called a person") and after him by a number of different authors from different schools. This aspect begins to be particularly prominent in the seventeenth century, but it is also linked to something that becomes even more decisive: a sense of subjective identity, that is, with the unity and continuity of the conscious life of the self. Locke (1690: bk 1, ch. 9, par. 29) defines a person as "a thinking intelligent being that has reason and reflection and can consider itself as itself, the same thinking thing, in different times and places". Even Kant, with all his insistence on rationality, claims that it is the fact of being able to represent to themselves their own selves that elevates persons infinitely above all the living beings on earth (Kant, 1775-80). The reference to self-consciousness and projectivity into the future, which is now widely accepted as a relevant factor in the concept, helps to clarify why, in contemporary bioethical debate, to say of some being that it is a person can mean not only to attribute a kind of value to that being, but also to ascribe the right to life to it, or to say that it is intrinsically wrong to destroy it. The idea is that to be harmed by the loss of its future, an entity should be aware of itself as having a future (Singer, 1993; Tooley, 1983). Though until recently most authors, following the widespread trend we hinted at above, more or less covertly identified this unified stream of consciousness with the human self, the rise of applied ethics had consequences in this field, too. Many moral philosophers who adopted something akin to this interpretation of person also applied the term to some nonhuman selves. The idea has appeared under other denominations, like that of "subject of a life" (Regan, 1983) or that of a being endowed with a biographical, rather than a merely biological, life (Rachels, 1986). This idea is central to *The Great Ape Project.*

3.   THE CASE FOR THE PERSONHOOD OF NONHUMAN APES

In *The Great Ape Project* we invited many scientists, philosophers, and others to write about the nature and moral status of our closest relatives. The theme of relatedness is widely present in scientific essays. The shaky nature of the boundary between human beings and the other great apes is emphasized by most authors: we share more than 98% of our DNA with chimpanzees, and only slightly less with gorillas and orangutans. As Robin Dunbar, professor of biological anthropology, points out, nonhuman great apes "differ only slightly more in their degree of genetic relatedness to you and me than do other populations of humans living elsewhere in the world" (Dunbar, 1993: 111-112). Jared Diamond (1992), whose expertise ranges from physiology to ethology, goes so far as to maintain that humans do not constitute a distinct family, nor even a distinct genus, but belong in the same genus as common chimpanzees and bonobos. Since our genus name was proposed first and takes priority, he suggests that there are today on earth three species in the genus *Homo*: the common chimpanzee, *Homo troglodytes*; the bonobo, *Homo paniscus*; and the third, or human chimpanzee, *Homo sapiens*. Relatedness is also present in Richard Dawkins's essay, in the form of an attack on the "discontinuous mind", which, by dividing animals up into discontinuous species, forgets that in the evolutionary view of life there must always be intermediates, and that it is sheer luck if they are no longer here to fill the gaps that the discontinuous mind erects (Dawkins, 1993). These and other contributions, like the one in which James Rachels (1990) directly addresses Darwin's theory, help to erase the idea of a sharp separation between us and the other great apes by removing its traditional background. They also contribute to explaining the second main element of *The Great Ape Project*, that is, similarity.

The most evident similarity is in linguistic capacity. Language has long been considered a human prerogative. In fact it still is, as far as it can be judged by the reactions to the recent studies in interspecies communication. Only prejudice can lead anyone to deny that language is being used in the Ameslan (or American Sign Language) interactions that occur between Francine Patterson and the gorillas Koko and Michael (Patterson and Gordon, this volume); between Lyn Miles and the orangutan Chantek; between Deborah and Roger Fouts and the community of chimpanzees associated with Washoe; and between Gary Shapiro and his orangutan "daughter" Princess, who lives in the wilds of Borneo. Language has always been linked with important mental capacities, and certainly the great apes' acquisition and use of Ameslan does point to such capacities, in particular for self-consciousness and for relating to others in sophisticated ways. Self-consciousness is most strikingly demonstrated by the fact that chimpanzees sometimes "think aloud", signing to themselves in appropriate ways when alone. As for the way in which chimpanzees relate to others, consider this: Washoe was looking at a magazine when her adopted son, Loulis (to whom she had spontaneously taught sign language), snatched the magazine and raced off out of the room. Washoe was left alone, signing "dirty, dirty". Chantek also knew the sign "dirty" and was taught to use it when he needed to go to the toilet, but he sometimes used it deceptively, in order to get into the bathroom to play with the soap and washing machine. As Miles underscores, in order to deceive, one must be able to see the events from the other's perspective and try to negate his or her perception. Koko, who has developed a vocabulary of over 1,000 signs, recognizes herself in front of a mirror, something that is considered an important indicator of self-awareness, and also enjoys making faces at herself and examining her teeth. In addition she makes jokes (and laughs at her own jokes) and refers appropriately to past events (Patterson and Gordon, this volume).

We could go on indefinitely with examples like these. But one episode can perhaps convey in a more immediate way what we share with the nonhuman apes. Geza Teleki, who spent much time in Africa studying free chimpanzees, was once sitting alone on the crest of a grassy ridge watching a sunset, when he noticed two adult male chimpanzees climbing toward him on opposite slopes. As they topped the crest, they saw one another and suddenly stood upright and advanced until they were face to face. Each extended his right hand to clasp and vigorously shake the other's. Then they sat down nearby and watched the sunset enfold the park.

On the basis of this evidence, it is surely indisputable that the notion of person can meaningfully be applied to the other great apes, and this is, in fact, what most authors do. At this point in the argument, this seems a natural claim. All the more troubling, then, is the extent to which it clashes with the real situation of the other apes. The other apes are currently forced to live and die in appalling ways because they are denied the basic protection owed to persons. There are chimpanzees in laboratories who spend their entire lives alone in small cages. They are infected with diseases like hepatitis or viruses like HIV and from that moment until they die are subjected to bleedings, biopsies, and laparatomies. What can this involve, not only in terms of physical suffering, but also for their inner life? Like the gorillas kept in zoos or the orangutans employed in entertainment, if they were born in captivity, they *know of no other reality*. If, on the other hand, they were captured in the African forests, their mothers were probably murdered before their eyes, their family and social ties were broken, and they arrived at their final overseas destination after a voyage that killed most of their fellow travelers. An answer to our question can come from Bimbo, a baby orangutan rescued from his wooden crate after having been shipped upside-down from Singapore to Bangkok. Despite his curious and vivacious character, and despite the assiduous care that improved his physical condition, Bimbo had lost his will to live. He stopped eating and allowed himself to die.

For those who hold that the case for ape equality is sound, the conclusion to draw must be political. It is evident that the case for equality for the other apes faces major difficulties, even if these are less dramatic than the difficulties that face the case for a more general change in the moral status of animals. Though attacked on many fronts at a theoretical level, humanism is still the ideology of our age. Considerable interests support the exploitation of the other apes. If we add that, as has often been underscored, nonhumans lack the ability to stand up in their own defense, the picture can seem unfavorable at the very least. However, there are past and contemporary experiences to draw on. First, there is the history of the abolition of the intra-human institution of slavery. In the last two hundred years, human slavery has been eliminated, or virtually eliminated, from the face of the earth. Along with it has gone the idea that human beings can be property. We need an international organization that can play for the liberation of the other apes the role played in the past by the Anti-Slavery Society. The goal of this organization will be to bring it to pass that all apes are removed from the category of property, both legally and morally, and are included in the category of persons. Once this goal is achieved, guardians will be needed to protect the newly sanctioned rights to life, liberty, and freedom from torture of the other apes. This is not an obstacle: it is exactly what already happens with immature or less gifted members of our species. Finally, even the idea of protected, independent territories where nonhuman apes could regain the dignity of an autonomous life is not without precedent, as the existence of human regions in need of external protection, the United Nations Trust Territories, shows.

We hope that this course of action will be supported by many who are involved in fields of expertise that cross the area of concern of the Great Ape Project. As scholars, they will thereby help to bridge the gap between the achievements of their disciplines and deeply rooted attitudes and paradigms. But, more important, they will follow a noble intellectual tradition of sparking change and contributing to make the world a better place in which to live.

REFERENCES

Aquinas, T., *Summa Theologiae,* I and III, Various editions.

Aristotle, *Nicomachean Ethics*, Various editions.

Boethius, *De Duabus Naturis et Una Persona Christi*, Various editions.

Cavalieri, P. and Singer, P. (Eds.), 1993, *The Great Ape Project: Equality beyond Humanity*, London: Fourth Estate.

Dawkins, R., 1993, Gaps in the mind. Pp. 80-87 in: (Eds. P. Cavalieri and P. Singer), *The Great Ape Project: Equality beyond Humanity*, London: Fourth Estate.

Descartes, R., *Discourse on Method*, Various editions.

Diamond, J., 1992, *The Third Chimpanzee*, New York: Harper Collins.

Dunbar, R.M., 1993, What's in a classification? Pp. 109-112 in: (Eds. P. Cavalieri and P. Singer), *The Great Ape Project: Equality beyond Humanity*, London: Fourth Estate.

Epictetus, *The Manual* and *Discourses*, Various editions.

Ferry, L., 1992, *Le Nouvel Ordre Écologique: L'arbre, L'animal et L'homme*, Paris: Grasset.

Kant, I., 1775-80, *Lectures on Ethics*, Various editions.

Kant, I., 1775-80, *Pragmatic Anthropology*, Various editions.

Locke, J., 1690, *Essay Concerning Human Understanding.*

MacIntyre, A., 1981, *After Virtue. A Study in Moral Theory*, Indiana: University of Notre Dame Press.

Patterson, F.G.P. and Gordon, W.L., this volume.

Rachels, J., 1986, *The End of Life*, Oxford: Oxford University Press

Rachels, J., 1990, *Created from Animals: The Moral Implications of Darwinism*, Oxford: Oxford University Press.

Regan, T., 1983, *The Case for Animal Rights*, Berkeley: University of California Press.

Sidgwick, H., 1907, *The Methods of Ethics*, 7th ed., London: Macmillan.

Silberbauer, G., 1991, Ethics in small-scale societies. In: (Ed. P. Singer), *A Companion to Ethics*, Oxford: Blackwell.

Singer, P., 1993, *Practical Ethics* (2nd ed.), Cambridge: Cambridge University Press.

Tooley, M., 1983, *Abortion and Infanticide,* Oxford: Oxford University Press.

Trendelenburg, A., 1910, A contribution to the history of the word person, *Monist* 20: 336-363.

Index

CPSIA information can be obtained
at www.ICGtesting.com
Printed in the USA
LVOW03s0241160118
563013LV00010B/205/P